QUANTUM MECHANICS: A MODERN APPROACH

Edited by
Deah E. Elmundo

Delve Publishing

Quantum Mechanics: A Modern Approach

Edited by Deah E. Elmundo

Edition 2018

Delve Publishing

Published by

Delve Publishing LLC

708 3rd Avenue, 6th Floor

New York

NY 10801

United States of America

www.delvepublishing.com

ISBN: 978-1-68095-806-5 (Hardback)

Printed in United States of America

Library of Congress Control Number: 2016920689

About the Editor

Deah E. Elmundo

Deah obtained her BS in Physics for Teachers from the Philippine Normal University where she graduated Magna cum laude and completed her academic requirements leading to MS in Teaching Physics from De La Salle University. She loves teaching Classical and Modern Physics and is presently a full-time science faculty of a junior high school in her country. She is also a maths and physics content specialist to a few tuition companies in Singapore and Australia

Table of Contents

2 QUANTUM THEORY FROM THE YEARS 1900–1925

3 FUNDAMENTALS OF QUANTUM MECHANICS

4 FORMULATIONS IN QUANTUM MECHANICS

5 EQUATIONS IN QUANTUM MECHANICS

6 MODERN INTERPRETATION OF QUANTUM MECHANICS

7 QUANTUM GRAVITY

8 RELATIVISTIC QUANTUM MECHANICS

List of Abbreviations

POVM	Positive-Operator Valued Measure
PPT	Positive Partial Transpose
QED	Quantum Electrodynamics
QFT	Quantum Field Theory
QG	Quantum Gravity
QM	Quantum Mechanics
RQM	Relativistic Quantum Mechanics
RWE	Relativistic Wave Equation
SED	Stochastic Electrodynamics
TOE	Theory of Everything
ZPF	Zero-Point Field

Preface

Quantum mechanics is the science of the very small. It explains the behavior of matter and its relations with energy on the scale of atoms and subatomic particles. Quantum mechanics (QM) developed over many decades, beginning as a set of controversial mathematical explanations of experiments that the math of classical mechanics could not explain. It began at the turn of the 20th century, around the same time that Albert Einstein published his theory of relativity, a separate mathematical revolution in physics that describes the motion of things at high speeds. Unlike relativity, however, the origins of QM cannot be attributed to any one scientist. Rather, multiple scientists contributed to a foundation of three revolutionary principles that gradually gained acceptance and experimental verification between 1900 and 1930. Light behaves in some respects like particles and in other respects like waves. Matter particles such as electrons and atoms exhibits wavelike behavior too. Some light sources, including neon lights, give off only certain frequencies of light. Quantum mechanics shows that light, along with all other forms of electromagnetic radiation, comes in discrete units, called photons, and predicts its energies, colors, and spectral intensities. Many aspects of quantum mechanics are counterintuitive and can seem paradoxical, because they describe behavior quite different from that seen at larger length scales. The principles of quantization, wave-particle duality and the uncertainty principle ushered in a new era for QM. In 1927, Paul Dirac applied a quantum understanding of electric and magnetic fields to give rise to the study of "quantum field theory" (QFT), which treated particles (such as photons and electrons) as excited states of an underlying physical field. Work in QFT continued for a decade until scientists hit a roadblock: Many equations in QFT stopped making physical sense because they produced results of infinity. After a decade of stagnation, Hans Bethe made a breakthrough in 1947 using a technique called "renormalization".

Content Coverage

Chapters One and *Two* hypothesize an introduction to quantum mechanics and old quantum theory, respectively. The old quantum theory yields a recipe for the quantization of the energy levels of the harmonic oscillator, which, when combined with the Boltzmann probability distribution of thermodynamics, yields the correct expression for the stored energy and specific heat of a quantum oscillator both at low and at ordinary temperatures.

Chapters Three and *Four* are intended to explore quantum mechanics concepts and formulations in quantum mechanics, separately. The particle and wave aspects of physical objects are such complementary phenomena. Both concepts are borrowed from classical mechanics, where it is impossible to be a particle and wave at the same time, whereas formulations in both classical and quantum mechanics provide exactly the same predictions and experimental results

Chapters Five presents exploration of equations in quantum mechanics given by renowned physicists such as Dirac, Klein-Gordon, Pauli, Rydberg, and Schrödinger.

Chapter Six is intended to give a modern interpretation of quantum mechanics, generalizing the conventional Copenhagen interpretation and providing a natural interpretation of quantum cosmology. This interpretation of quantum mechanics is based on a consistency criterion that then allows probabilities to be assigned to various alternative histories of a system such that the probabilities for each history obey the rules of classical probability while being consistent with the Schrödinger equation.

Chapters Seven and *Eight* clarify the fundamentals of quantum gravity and relativistic quantum mechanics, separately. The relativistic formulation is more successful than the original quantum mechanics in some contexts, in particular: the prediction of antimatter, electron spin, spin magnetic moments of elementary spin-1/2 fermions, fine structure, and quantum dynamics of charged particles in electromagnetic fields.

CHAPTER 1

Introduction to Quantum Mechanics

Chapter Contents

The term quantum mechanics (QM) generally refers to a mathematical formulation of quantum theory. It often describes and predicts the movement and behavior of particles at the atomic and subatomic levels. These include particles such as atoms, electrons, protons, and photons. According to QM, the behavior and movement of particles at these microscopic levels are counter-intuitive, radically differing from anything observed in everyday life. This requires a theory like quantum mechanics in order to better explain this different world.

INTRODUCTION

Quantum mechanics is the science of the very small. It explains the behaviour of matter and its interactions with energy on the scale of atoms and subatomic particles.

By contrast, classical physics only explains matter and energy on a scale familiar to human experience, including the behaviour of astronomical bodies such as the Moon. Classical physics is still used in much of modern science and technology. However, towards the end of the 19th century, scientists discovered phenomena in both the large (macro) and the small (micro) worlds that classical physics could not explain. Coming to terms with these limitations led to two major revolutions in physics which created a shift in the original scientific paradigm: the theory of relativity and the development of quantum mechanics. This article describes how physicists discovered the limitations of classical physics and developed the main concepts of the quantum theory that replaced it in the early decades of the 20th century. These concepts are described in roughly the order in which they were first discovered. For a more complete history of the subject, see History of quantum mechanics.

Light behaves in some respects like particles and in other respects like waves. Matter—particles such as electrons and atoms—exhibits wavelike behaviour too. Some light sources, including neon lights, give off only certain frequencies of light. Quantum mechanics shows that light, along with all other forms of electromagnetic radiation, comes in discrete units, called photons, and predicts its energies, colours, and spectral intensities. Since one never observes half a photon, a single photon is a quantum, or smallest observable amount, of the electromagnetic field. More broadly, quantum mechanics shows that many quantities, such as angular momentum, that appeared to be continuous in the zoomed-out view of classical mechanics, turn out to be (at the small, zoomed-in scale of quantum mechanics) quantized. Angular momentum is required to take on one of a set of discrete allowable values, and since the gap between these values is so minute, the discontinuity is only apparent at the atomic level.

Many aspects of quantum mechanics are counterintuitive and can seem paradoxical, because they describe behaviour quite different from that seen at larger length scales. In the words of quantum physicist Richard Feynman, quantum mechanics deals with "nature as She is – absurd". For example, the uncertainty principle of quantum mechanics means that the more closely one pins down one measurement (such as the position of a particle), the less accurate another measurement pertaining to the same particle (such as its momentum) must become.

THE FIRST QUANTUM THEORY: MAX PLANCK AND BLACK-BODY RADIATION

Thermal radiation is electromagnetic radiation emitted from the surface of an object due to the object's internal energy. If an object is heated sufficiently, it starts to emit light at the red end of the spectrum, as it becomes red hot.

Heating it further causes the colour to change from red to yellow, white, and blue, as light at shorter wavelengths (higher frequencies) begins to be emitted. A perfect emitter is also a perfect absorber: when it is cold, such an object looks perfectly black, because it absorbs all the light that falls on it and emits none. Consequently, an ideal thermal emitter is known as a black body, and the radiation it emits is called black-body radiation.

In the late 19th century, thermal radiation had been fairly well characterized experimentally. However, classical physics led to the Rayleigh-Jeans law, which, as shown in the figure, agrees with experimental results well at low frequencies, but strongly disagrees at high frequencies. Physicists searched for a single theory that explained all the experimental results.

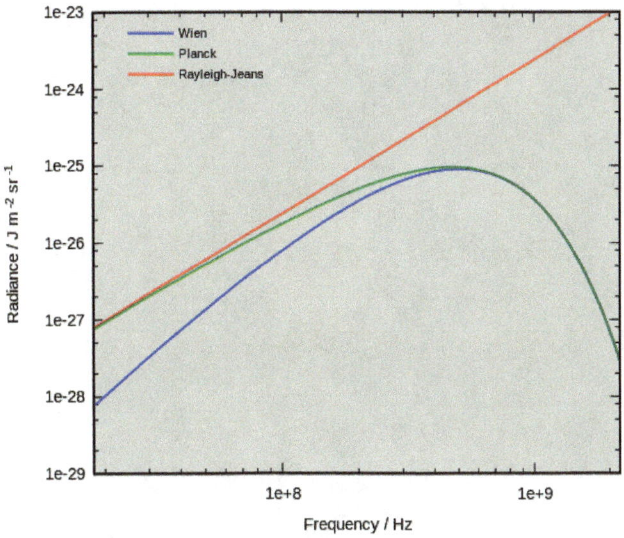

Predictions of the amount of thermal radiation of different frequencies emitted by a body. Correct values predicted by Planck's law (green) contrasted against the classical values of Rayleigh-Jeans law (red) and Wien approximation (blue).

The first model that was able to explain the full spectrum of thermal radiation was put forward by Max Planck in 1900. He proposed a mathematical model in which the thermal radiation was in equilibrium with a set of harmonic oscillators. To reproduce the experimental results, he had to assume that each oscillator emitted an integer number of units of energy at its single characteristic frequency, rather than being able to emit any arbitrary amount of energy. In other words, the energy emitted by an oscillator was quantized. The quantum of energy for each oscillator, according to Planck, was proportional to the frequency of the oscillator; the constant of proportionality is now known as the Planck constant. The Planck constant, usually written as h, has the value of 6.63×10^{-34} J s. So, the energy E of an oscillator of frequency f is given by

$$E = nhf, \quad \text{where} \quad n = 1, 2, 3, \ldots$$

To change the colour of such a radiating body, it is necessary to change its temperature. Planck's law explains why: increasing the temperature of a body allows it to emit more energy overall, and means that a larger proportion of the energy is towards the violet end of the spectrum.

Planck's law was the first quantum theory in physics, and Planck won the Nobel Prize in 1918 "in recognition of the services he rendered to the advancement of Physics by his discovery of energy quanta". At the time, however, Planck's view was that quantization was purely a heuristic mathematical construct, rather than (as is now believed) a fundamental change in our understanding of the world.

Hot metalwork. The yellow-orange glow is the visible part of the thermal radiation emitted due to the high temperature. Everything else in the picture is glowing with thermal radiation as well, but less brightly and at longer wavelengths than the human eye can detect. A far-infrared camera can observe this radiation.

Photons: The Quantisation of Light

Albert Einstein circa 1905.

In 1905, Albert Einstein took an extra step. He suggested that quantisation was not just a mathematical construct, but that the energy in a beam of light actually occurs in individual packets, which are now called photons. The energy of a single photon is given by its frequency multiplied by Planck's constant:

$$E = hf$$

For centuries, scientists had debated between two possible theories of light: was it a wave or did it instead comprise a stream of tiny particles? By the 19th century, the debate was generally considered to have been settled in favour of the wave theory, as it was able to explain observed effects such as refraction, diffraction, interference and polarization. James Clerk Maxwell had shown that electricity, magnetism and light are all manifestations of the same phenomenon: the electromagnetic field. Maxwell's equations, which are the complete set of laws of classical electromagnetism, describe light as waves: a combination of oscillating electric and magnetic fields. Because of the preponderance of evidence in favour of the wave theory, Einstein's ideas were met initially with great skepticism. Eventually, however, the photon model became favoured. One of the most significant pieces of evidence in its favour was its ability to explain several puzzling properties of the photoelectric effect, described in the following section. Nonetheless, the wave analogy remained indispensable for helping to understand other characteristics of light: diffraction, refraction and interference.

The Photoelectric Effect

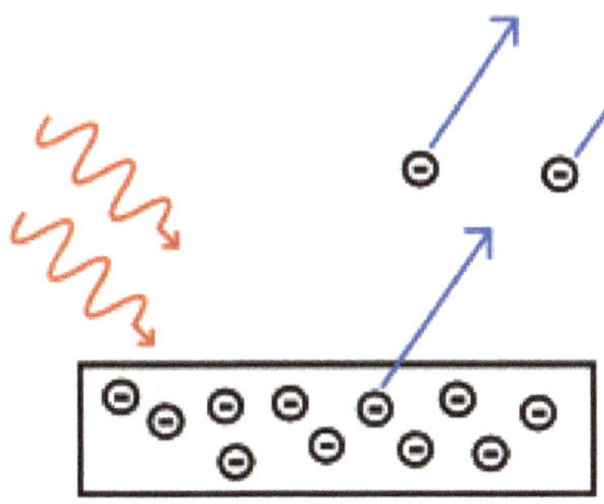

Light (red arrows, left) is shone upon a metal. If the light is of sufficient frequency (i.e. sufficient energy), electrons are ejected (blue arrows, right).

In 1887, Heinrich Hertz observed that when light with sufficient frequency hits a metallic surface, it emits electrons. In 1902, Philipp Lenard discovered that the maximum possible energy of an ejected electron is related to the frequency of the light, not to its intensity: if the frequency is too low, no electrons are ejected regardless of the intensity. Strong beams of light toward the red end of the spectrum might produce no electrical potential at all, while weak beams of light toward the violet end of the spectrum would produce higher and higher voltages. The lowest frequency of light that can cause electrons to be emitted, called the threshold frequency, is different for different metals. This observation is at odds with classical electromagnetism, which predicts that the electron's energy should be proportional to the intensity of the radiation.[24] So when physicists first discovered devices exhibiting the photoelectric effect, they initially expected that a higher intensity of light would produce a higher voltage from the photoelectric device.

Einstein explained the effect by postulating that a beam of light is a stream of particles (photons) and that, if the beam is of frequency f, then each photon has an energy equal to hf. An electron is likely to be struck only by a single photon, which imparts at most an energy hf to the electron. Therefore, the intensity of the beam has no effect and only its frequency determines the maximum energy that can be imparted to the electron.

To explain the threshold effect, Einstein argued that it takes a certain amount of energy, called the work function and denoted by φ, to remove an electron from the metal. This amount of energy is different for each metal. If the energy of the photon is less than the work function, then it does not carry sufficient energy to remove the electron from the metal. The threshold frequency, f_0, is the frequency of a photon whose energy is equal to the work function:

$$\varphi = hf_0.$$

If f is greater than f_0, the energy hf is enough to remove an electron. The ejected electron has a kinetic energy, E_K, which is, at most, equal to the photon's energy minus the energy needed to dislodge the electron from the metal:

$$E_K = hf - \varphi = h(f - f_0).$$

Einstein's description of light as being composed of particles extended Planck's notion of quantised energy, which is that a single photon of a given frequency, f, delivers an invariant amount of energy, hf. In other words, individual photons can deliver more or less energy, but only depending on their frequencies. In nature, single photons are rarely encountered. The Sun and emission sources available in the 19th century emit vast numbers of photons every second, and so the importance of the energy carried by each individual photon was not obvious. Einstein's idea that the energy contained in individual units of light depends on their frequency made it possible to explain experimental results that had hitherto seemed quite counterintuitive. However, although the photon is a particle, it was still being described as having the wave-like property of frequency. Once again, the particle account of light was being compromised.

Consequences of the Light Being Quantised

The relationship between the frequency of electromagnetic radiation and the energy of each individual photon is why ultraviolet light can cause sunburn, but visible or infrared light cannot. A photon of ultraviolet light will deliver a high amount of energy – enough to contribute to cellular damage such as occurs in a sunburn. A photon of infrared light will deliver a lower amount of energy – only enough to warm one's skin. So, an infrared lamp can warm a large surface, perhaps large enough to keep people comfortable in a cold room, but it cannot give anyone a sunburn.

All photons of the same frequency have identical energy, and all photons of different frequencies have proportionally (order 1, $E_{photon} = hf$) different energies. However, although the energy imparted by photons is invariant at any given frequency, the initial energy state of the electrons in a photoelectric device prior to absorption of light is not necessarily uniform. Anomalous results may occur in the case of individual electrons. For instance, an electron that was already excited above the equilibrium level of the photoelectric device might be ejected when it absorbed uncharacteristically low frequency illumination. Statistically, however, the characteristic behaviour of a photoelectric device will reflect the behaviour of the vast majority of its electrons, which will be at their equilibrium level. This point is helpful in comprehending the distinction between the study of individual particles in quantum dynamics and the study of massed particles in classical physics.

THE QUANTISATION OF MATTER: THE BOHR MODEL OF THE ATOM

By the dawn of the 20th century, evidence required a model of the atom with a diffuse cloud of negatively charged electrons surrounding a small, dense, positively charged nucleus. These properties suggested a model in which the electrons circle around the nucleus like planets orbiting a sun. However, it was also known that the atom in this model would be unstable: according to classical theory, orbiting electrons are undergoing centripetal acceleration, and should therefore give off electromagnetic radiation, the loss of energy also causing them to spiral toward the nucleus, colliding with it in a fraction of a second.

A second, related, puzzle was the emission spectrum of atoms. When a gas is heated, it gives off light only at discrete frequencies. For example, the visible light given off by hydrogen consists of four different colours, as shown in the picture below. The intensity of the light at different frequencies is also different. By contrast, white light consists of a continuous emission across the whole range of visible frequencies. By the end of the nineteenth century, a simple rule known as Balmer's formula had been found which showed how the frequencies of the different lines were related to each other, though without explaining why this was, or making any prediction about the intensities. The formula also predicted some additional spectral lines in ultraviolet and infrared light which had not been observed at the time. These lines were later observed experimentally, raising confidence in the value of the formula.

Emission spectrum of hydrogen. When excited, hydrogen gas gives off light in four distinct colours (spectral lines) in the visible spectrum, as well as a number of lines in the infrared and ultraviolet.

The Mathematical Formula Describing Hydrogen's Emission Spectrum

In 1885 the Swiss mathematician Johann Balmer discovered that each wavelength λ (lambda) in the visible spectrum of hydrogen is related to some integer n by the equation

$$\lambda = B\left(\frac{n^2}{n^2-4}\right) \qquad n = 3, 4, 5, 6$$

where B is a constant which Balmer determined to be equal to 364.56 nm.

In 1888 Johannes Rydberg generalized and greatly increased the explanatory utility of Balmer's formula. He predicted that λ is related to two integers n and maccording to what is now known as the Rydberg formula:

$$\frac{1}{\lambda} = R\left(\frac{1}{m^2} - \frac{1}{n^2}\right),$$

where R is the Rydberg constant, equal to 0.0110 nm^{-1}, and n must be greater than m.

Rydberg's formula accounts for the four visible wavelengths of hydrogen by setting m = 2 and n = 3, 4, 5, 6. It also predicts additional wavelengths in the emission spectrum: for m = 1 and for n > 1, the emission spectrum should contain certain ultraviolet wavelengths, and for m = 3 and n > 3, it should also contain certain infrared wavelengths. Experimental observation of these wavelengths came two decades later: in 1908 Louis Paschen found some of the predicted infrared wavelengths, and in 1914 Theodore Lyman found some of the predicted ultraviolet wavelengths.

Note that both Balmer and Rydberg's formulas involve integers: in modern terms, they imply that some property of the atom is quantised. Understanding exactly what this property was, and why it was quantised, was a major part in the development of quantum mechanics, as will be shown in the rest of this article.

In 1913 Niels Bohr proposed a new model of the atom that included quantized electron orbits: electrons still orbit the nucleus much as planets orbit around the sun, but they are only permitted to inhabit certain orbits, not to orbit at any distance. When an atom emitted (or absorbed) energy, the electron did not move in a continuous trajectory from one orbit around the nucleus to another, as might be expected classically. Instead, the electron would jump instantaneously from one orbit to another, giving off the emitted light in the form of a photon. The possible energies of photons given off by each element were determined by the differences in energy between the orbits, and so the emission spectrum for each element would contain a number of lines.

Niels Bohr as a young man

Starting from only one simple assumption about the rule that the orbits must obey, the Bohr model was able to relate the observed spectral lines in the emission spectrum of hydrogen to previously known constants. In Bohr's model the electron simply wasn't allowed to emit energy continuously and crash into the nucleus: once it was in the closest permitted orbit, it was stable forever. Bohr's model didn't explain why the orbits should be quantised in that way, nor was it able to make accurate predictions for atoms with more than one electron, or to explain why some spectral lines are brighter than others.

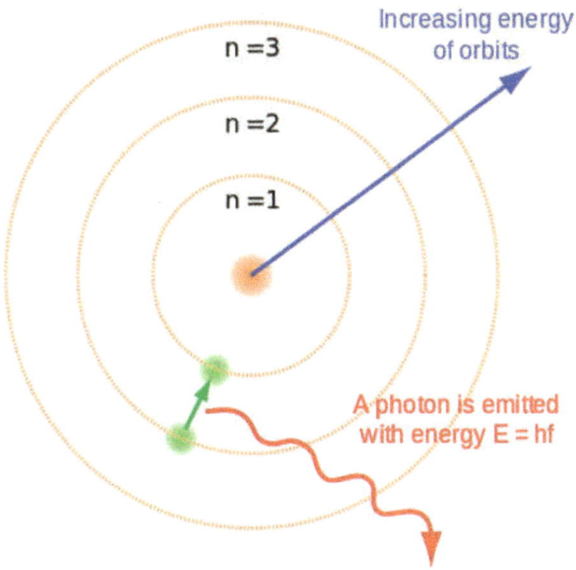

The Bohr model of the atom, showing an electron transitioning from one orbit to another by emitting a photon.

Although some of the fundamental assumptions of the Bohr model were soon found to be wrong, the key result that the discrete lines in emission spectra are due to some property of the electrons in atoms being quantised is correct. The way that the electrons actually behave is strikingly different from Bohr's atom, and from what we see in the world of our everyday experience; this modern quantum mechanical model of the atom is discussed below.

A more detailed explanation of the Bohr model

Bohr theorised that the angular momentum, L, of an electron is quantised:

$$L = n\frac{h}{2\pi} = n\hbar$$

where n is an integer and h is the Planck constant. Starting from this assumption, Coulomb's law and the equations of circular motion show that an electron with n units of angular momentum will orbit a proton at a distance r given by

$$r = \frac{n^2h^2}{4\pi^2 k_e m e^2},$$

where k_e is the Coulomb constant, m is the mass of an electron, and e is the charge on an electron. For simplicity this is written as

$$r = n^2 a_0,$$

where a_0, called the Bohr radius, is equal to 0.0529 nm. The Bohr radius is the radius of the smallest allowed orbit.

The energy of the electron can also be calculated, and is given by

$$E = -\frac{k_e e^2}{2a_0}\frac{1}{n^2}$$

Thus Bohr's assumption that angular momentum is quantised means that an electron can only inhabit certain orbits around the nucleus, and that it can have only certain energies. A consequence of these constraints is that the electron will not crash into the nucleus: it cannot continuously emit energy, and it cannot come closer to the nucleus than a_0 (the Bohr radius).

An electron loses energy by jumping instantaneously from its original orbit to a lower orbit; the extra energy is emitted in the form of a photon. Conversely, an electron that absorbs a photon gains energy, hence it jumps to an orbit that is farther from the nucleus.

Each photon from glowing atomic hydrogen is due to an electron moving from a higher orbit, with radius r_n, to a lower orbit, r_m. The energy E_γ of this photon is the difference in the energies E_n and E_m of the electron:

$$E_\gamma = E_n - E_m = \frac{k_e e^2}{2a_0}\left(\frac{1}{m^2} - \frac{1}{n^2}\right)$$

Since Planck›s equation shows that the photon›s energy is related to its wavelength by $E_\gamma = hc/\lambda$, the wavelengths of light that can be emitted are given by

$$\frac{1}{\lambda} = \frac{k_e e^2}{2a_0 hc}\left(\frac{1}{m^2} - \frac{1}{n^2}\right).$$

This equation has the same form as the Rydberg formula, and predicts that the constant R should be given by

$$R = \frac{k_e e^2}{2a_0 hc}.$$

Therefore, the Bohr model of the atom can predict the emission spectrum of hydrogen in terms of fundamental constants. However, it was not able to make accurate predictions for multi-electron atoms, or to explain why some spectral lines are brighter than others.

WAVE-PARTICLE DUALITY

Louis de Broglie in 1929. De Broglie won the Nobel Prize in Physics for his prediction that matter acts as a wave, made in his 1924 PhD thesis.

Just as light has both wave-like and particle-like properties, matter also has wave-like properties.

Matter behaving as a wave was first demonstrated experimentally for electrons: a beam of electrons can exhibit diffraction, just like a beam of light or a water wave. Similar wave-like phenomena were later shown for atoms and even small molecules.

The wavelength, λ, associated with any object is related to its momentum, p, through the Planck constant, h:

$$p = \frac{h}{\lambda}.$$

The relationship, called the de Broglie hypothesis, holds for all types of matter: all matter exhibits properties of both particles and waves.

The concept of wave–particle duality says that neither the classical concept of "particle" nor of "wave" can fully describe the behaviour of quantum-scale objects, either photons or matter. Wave–particle duality is an example of the principle of complementarity in quantum physics. An elegant example of wave–particle duality, the double slit experiment, is discussed in the section below.

The Double-Slit Experiment

In the double-slit experiment, as originally performed by Thomas Young and Augustin Fresnel in 1827, a beam of light is directed through two narrow, closely spaced slits, producing an interference pattern of light and dark bands on a screen. If one of the slits is covered up, one might naively expect that the intensity of the fringes due to interference would be halved everywhere. In fact, a much simpler pattern is seen, a simple diffraction pattern. Closing one slit results in a much simpler pattern diametrically opposite the open slit. Exactly the same behaviour can be demonstrated in water waves, and so the double-slit experiment was seen as a demonstration of the wave nature of light.

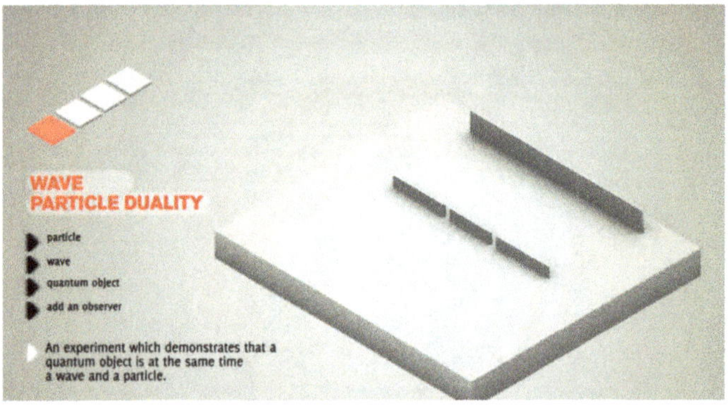

The double slit experiment for a classical particle, a wave, and a quantum particle demonstrating wave-particle duality

The double-slit experiment has also been performed using electrons, atoms, and even molecules, and the same type of interference pattern is seen. Thus it has been demonstrated that all matter possesses both particle and wave characteristics.

Even if the source intensity is turned down, so that only one particle (e.g. photon or electron) is passing through the apparatus at a time, the same interference pattern develops over time. The quantum particle acts as a wave when passing through the double slits, but as a particle when it is detected. This is a typical feature of quantum complementarity: a quantum particle will act as a wave in an experiment to measure its wave-like properties, and like a particle in an experiment to measure its particle-like properties. The point on the detector screen where any individual particle shows up will be the result of a random process. However, the distribution pattern of many individual particles will mimic the diffraction pattern produced by waves.

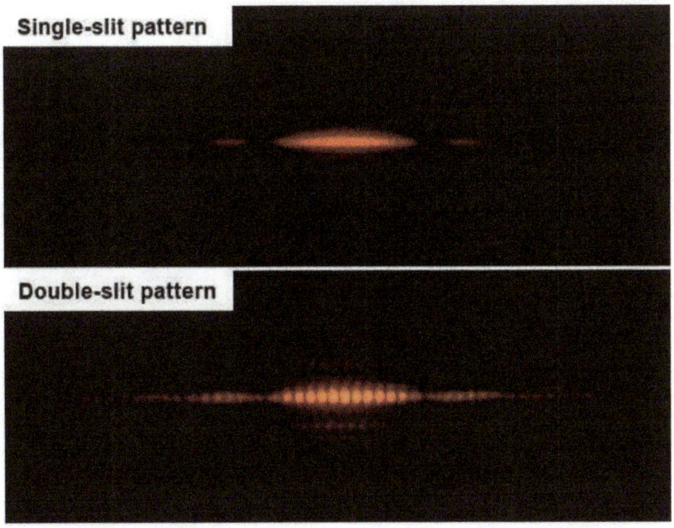

The diffraction pattern produced when light is shone through one slit (top) and the interference pattern produced by two slits (bottom). The much more complex pattern from two slits, with its small-scale interference fringes, demonstrates the wave-like propagation of light.

Application to the Bohr model

De Broglie expanded the Bohr model of the atom by showing that an electron in orbit around a nucleus could be thought of as having wave-like properties. In particular, an electron will be observed only in situations that permit a standing wave around a nucleus. An example of a standing wave is a violin string, which is fixed at both ends and can be made to vibrate. The waves created by a stringed instrument appear to oscillate in place, moving from crest to trough in an up-and-down motion. The wavelength of a standing wave is related to the length of the vibrating object and the boundary conditions. For example, because the violin string is fixed at both ends, it can carry standing waves of wavelengths $2l/n$, where l is the length and n is a positive integer. De Broglie suggested that the allowed electron orbits were those for which the circumference of the orbit would be an integer number of wavelengths. The electron's wavelength therefore determines that only Bohr orbits of certain distances from the nucleus are possible. In turn, at any distance from the nucleus smaller than a certain value it would be impossible to establish an orbit. The minimum possible distance from the nucleus is called the Bohr radius.

De Broglie's treatment of quantum events served as a starting point for Schrödinger when he set out to construct a wave equation to describe quantum theoretical events.

Spin

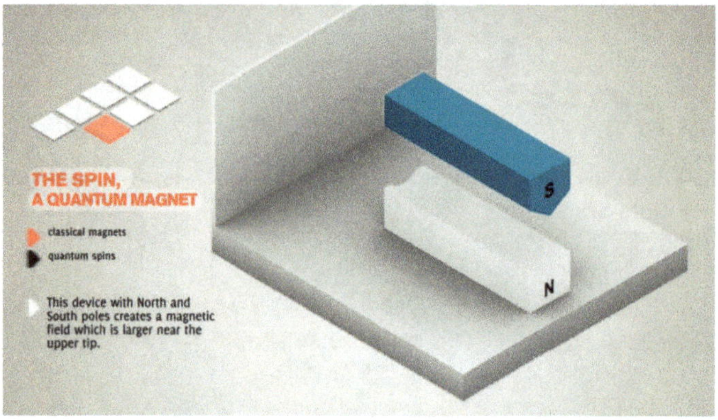

Quantum spin versus classical magnet in the Stern-Gerlach experiment.

In 1922, Otto Stern and Walther Gerlach shot silver atoms through an (inhomogeneous) magnetic field. In classical mechanics, a magnet thrown through a magnetic field may be, depending on its orientation (if it is pointing with its northern pole upwards or down, or somewhere in between), deflected a small or large distance upwards or downwards. The atoms that Stern and Gerlach shot through the magnetic field acted in a similar way. However, while the magnets could be deflected variable distances, the atoms would always be deflected a constant distance either up or down. This implied that the property of the atom which corresponds to the magnet's orientation must be quantised, taking one of two values (either up or down), as opposed to being chosen freely from any angle.

Ralph Kronig originated the idea that particles such as atoms or electrons behave as if they rotate, or "spin", about an axis. Spin would account for the missing magnetic moment, and allow two electrons in the same orbital to occupy distinct quantum states if they "spun" in opposite directions, thus satisfying the exclusion principle. The quantum number represented the sense (positive or negative) of spin.

The choice of orientation of the magnetic field used in the Stern-Gerlach experiment is arbitrary. In the animation shown here, the field is vertical and so the atoms are deflected either up or down. If the magnet is rotated a quarter turn, the atoms will be deflected either left or right. Using a vertical field shows that the spin along the vertical axis is quantised, and using a horizontal field shows that the spin along the horizontal axis is quantised.

If, instead of hitting a detector screen, one of the beams of atoms coming out of the Stern-Gerlach apparatus is passed into another (inhomogeneous) magnetic field oriented in the same direction, all of the atoms will be deflected the same way in this second field. However, if the second field is oriented at 90° to the first, then half of the atoms will be deflected one way and half the other, so that the atom's spin about the horizontal and vertical axes are independent of each other. However, if one of these beams (e.g. the atoms that were deflected up then left) is passed into a third magnetic field, oriented the same way as the first, half of the atoms will go one way and half the other, even though they all went in the same direction originally. The action of measuring the atoms' spin with respect to a horizontal field has changed their spin with respect to a vertical field.

The Stern-Gerlach experiment demonstrates a number of important features of quantum mechanics:

- a feature of the natural world has been demonstrated to be quantised, and only able to take certain discrete values
- particles possess an intrinsic angular momentum that is closely analogous to the angular momentum of a classically spinning object
- measurement changes the system being measured in quantum mechanics. Only the spin of an object in one direction can be known, and observing the spin in another direction will destroy the original information about the spin.
- quantum mechanics is probabilistic: whether the spin of any individual atom sent into the apparatus is positive or negative is random.

Development of Modern Quantum Mechanics

In 1925, Werner Heisenberg attempted to solve one of the problems that the Bohr model left unanswered, explaining the intensities of the different lines in the hydrogen emission spectrum. Through a series of mathematical analogies, he wrote out the quantum mechanical analogue for the classical computation of intensities. Shortly afterwards, Heisenberg's colleague Max Born realised that Heisenberg's method of calculating the probabilities for transitions between the different energy levels could best be expressed by using the mathematical concept of matrices.

In the same year, building on de Broglie's hypothesis, Erwin Schrödinger developed the equation that describes the behaviour of a quantum mechanical wave. The mathematical model, called the Schrödinger equation after its creator, is central to quantum mechanics, defines the permitted stationary states of a quantum system, and describes how the quantum state of a physical system changes in time. The wave itself is described by a mathematical function known as a "wave function". Schrödinger said that the wave function provides the "means for predicting probability of measurement results".

Schrödinger was able to calculate the energy levels of hydrogen by treating a hydrogen atom's electron as a classical wave, moving in a well of electrical potential created by the proton. This calculation accurately reproduced the energy levels of the Bohr model.

In May 1926, Schrödinger proved that Heisenberg's matrix mechanics and his own wave mechanics made the same predictions about the properties and behaviour of the electron; mathematically, the two theories had an underlying common form. Yet the two men disagreed on the interpretation of their mutual theory. For instance, Heisenberg accepted the theoretical prediction of jumps of electrons between orbitals in an atom, but Schrödinger hoped that a theory based on continuous wave-like properties could avoid what he called (as paraphrased by Wilhelm Wien) this nonsense about quantum jumps.

Copenhagen Interpretation

Bohr, Heisenberg and others tried to explain what these experimental results and mathematical models really mean. Their description, known as the Copenhagen interpretation of quantum mechanics, aimed to describe the nature of reality that was being probed by the measurements and described by the mathematical formulations of quantum mechanics.

The main principles of the Copenhagen interpretation are:

- A system is completely described by a wave function, usually represented by the Greek letter ψ ("psi"). (Heisenberg)
- How ψ changes over time is given by the Schrödinger equation.
- The description of nature is essentially probabilistic. The probability of an event – for example, where on the screen a particle will show up in the two slit experiment – is related to the square of the absolute value of the amplitude of its wave function. (Born rule, due to Max Born, which gives a physical meaning to the wave function in the Copenhagen interpretation: the probability amplitude)
- It is not possible to know the values of all of the properties of the system at the same time; those properties that are not known with precision must be described by probabilities. (Heisenberg's uncertainty principle)
- Matter, like energy, exhibits a wave–particle duality. An experiment can demonstrate the particle-like properties of matter, or its wave-like properties; but not both at the same time. (Complementarity principle due to Bohr)
- Measuring devices are essentially classical devices, and measure classical properties such as position and momentum.
- The quantum mechanical description of large systems should closely approximate the classical description. (Correspondence principle of Bohr and Heisenberg)

Various consequences of these principles are discussed in more detail in the following subsections.

Uncertainty Principle

Suppose it is desired to measure the position and speed of an object – for example a car going through a radar speed trap. It can be assumed that the car has a definite position and speed at a particular moment in time. How accurately these values can be measured depends on the quality of the measuring equipment – if the precision of the measuring equipment is improved, it will provide a result that is closer to the true value. It might be assumed that the speed of the car and its position could be operationally defined and measured simultaneously, as precisely as might be desired.

Werner Heisenberg at the age of 26. Heisenberg won the Nobel Prize in Physics in 1932 for the work that he did at around this time.

The Niels Bohr Institute in Copenhagen, which was a focal point for researchers in quantum mechanics and related subjects in the 1920s and 1930s. Most of the world's best known theoretical physicists spent time there.

In 1927, Heisenberg proved that this last assumption is not correct. Quantum mechanics shows that certain pairs of physical properties, such as for example position and speed, cannot be simultaneously measured, nor defined in operational terms, to arbitrary precision: the more precisely one property is measured, or defined in operational terms, the less precisely can the other. This statement is known as the uncertainty principle. The uncertainty principle isn't only a statement about the accuracy of our measuring equipment, but, more deeply, is about the conceptual nature of the measured quantities – the assumption that the car had simultaneously defined position and speed does not work in quantum mechanics. On a scale of cars and people, these uncertainties are negligible, but when dealing with atoms and electrons they become critical.

Heisenberg gave, as an illustration, the measurement of the position and momentum of an electron using a photon of light. In measuring the electron's position, the higher the frequency of the photon, the more accurate is the measurement of the position of the impact of the photon with the electron, but the greater is the disturbance of the electron. This is because from the impact with the photon, the electron absorbs a random amount of energy, rendering the measurement obtained of its momentum increasingly uncertain (momentum is velocity multiplied by mass), for one is necessarily measuring its post-impact disturbed momentum from the collision products and not its original momentum. With a photon of lower frequency, the disturbance (and hence uncertainty) in the momentum is less, but so is the accuracy of the measurement of the position of the impact.

The uncertainty principle shows mathematically that the product of the uncertainty in the position and momentum of a particle (momentum is velocity multiplied by mass) could never be less than a certain value, and that this value is related to Planck's constant.

Wave Function Collapse

Wave function collapse is a forced expression for whatever just happened when it becomes appropriate to replace the description of an uncertain state of a system by a description of the system in a definite state. Explanations for the nature of the process of becoming certain are controversial. At any time before a photon "shows up" on a detection screen it can only be described by a set of probabilities for where it might show up. When it does show up, for instance in the CCD of an electronic camera, the time and the space where it interacted with the device are known within very tight limits. However, the photon has disappeared, and the wave function has disappeared with it. In its place some physical change in the detection screen has appeared, e.g., an exposed spot in a sheet of photographic film, or a change in electric potential in some cell of a CCD.

Eigenstates and Eigenvalues

Because of the uncertainty principle, statements about both the position and momentum of particles can only assign a probability that the position or momentum will have some numerical value. Therefore, it is necessary to formulate clearly the difference between the state of something that is indeterminate, such as an electron in a probability cloud, and the state of something having a definite value. When an object can definitely be "pinned-down" in some respect, it is said to possess an eigenstate.

In the Stern-Gerlach experiment discussed above, the spin of the atom about the vertical axis has two eigenstates: up and down. Before measuring it, we can only say that any individual atom has equal probability of being found to have spin up or spin down. The measurement process causes the wavefunction to collapse into one of the two states.

The eigenstates of spin about the vertical axis are not simultaneously eigenstates of spin about the horizontal axis, so this atom has equal probability of being found to have either value of spin about the horizontal axis. As described in the section above, measuring the spin about the horizontal axis can allow an atom which was spin up to become spin down: measuring its spin about the horizontal axis collapses its wave function into one of the eigenstates of this measurement, which means it is no longer in an eigenstate of spin about the vertical axis, so can take either value.

The Pauli Exclusion Principle

Wolfgang Pauli

In 1924, Wolfgang Pauli proposed a new quantum degree of freedom (or quantum number), with two possible values, to resolve inconsistencies between observed molecular spectra and the predictions of quantum mechanics. In particular, the spectrum of atomic hydrogen had a doublet, or pair of lines differing by a small amount, where only one line was expected. Pauli formulated his exclusion principle, stating that "There cannot exist an atom in such a quantum state that two electrons within it have the same set of quantum numbers."

A year later, Uhlenbeck and Goudsmit identified Pauli's new degree of freedom with the property called spin whose effects were observed in the Stern–Gerlach experiment.

Application to the Hydrogen Atom

Bohr's model of the atom was essentially a planetary one, with the electrons orbiting around the nuclear "sun." However, the uncertainty principle states that an electron cannot simultaneously have an exact location and velocity in the way that a planet does. Instead of classical orbits, electrons are said to inhabit atomic orbitals. An orbital is the "cloud" of possible locations in which an electron might be found, a distribution of probabilities rather than a precise location. Each orbital is three dimensional, rather than the two dimensional orbit, and is often depicted as a three-dimensional region within which there is a 95 percent probability of finding the electron.

Schrödinger was able to calculate the energy levels of hydrogen by treating a hydrogen atom's electron as a wave, represented by the "wave function" Ψ, in an electric potential well, V, created by the proton. The solutions to Schrödinger's equation are distributions of probabilities for electron positions and locations. Orbitals have a range of different shapes in three dimensions. The energies of the different orbitals can be calculated, and they accurately match the energy levels of the Bohr model.

Within Schrödinger's picture, each electron has four properties:

- An "orbital" designation, indicating whether the particle wave is one that is closer to the nucleus with less energy or one that is farther from the nucleus with more energy;
- The "shape" of the orbital, spherical or otherwise;
- The "inclination" of the orbital, determining the magnetic moment of the orbital around the z-axis.
- The "spin" of the electron.

The collective name for these properties is the quantum state of the electron. The quantum state can be described by giving a number to each of these properties; these are known as the electron's quantum numbers. The quantum state of the electron is described by its wave function. The Pauli exclusion principle demands that no two electrons within an atom may have the same values of all four numbers.

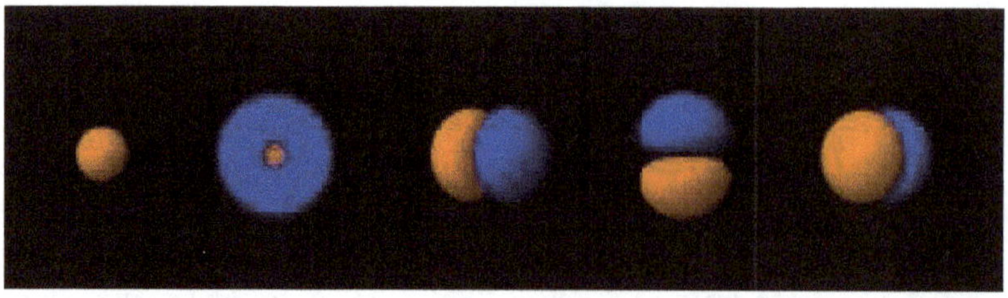

The shapes of the first five atomic orbitals: 1s, 2s, $2p_x$, $2p_y$, and $2p_z$. The colours show the phase of the wave function.

The first property describing the orbital is the principal quantum number, n, which is the same as in Bohr's model. n denotes the energy level of each orbital. The possible values for n are integers:

$$n = 1, 2, 3 \ldots$$

The next quantum number, the azimuthal quantum number, denoted l, describes the shape of the orbital. The shape is a consequence of the angular momentum of the orbital. The angular momentum represents the resistance of a spinning object to speeding up or slowing down under the influence of external force. The azimuthal quantum number represents the orbital angular momentum of an electron around its nucleus. The possible values for l are integers from 0 to n – 1 (where n is the principal quantum number of the electron):

$$l = 0, 1, \ldots, n - 1.$$

The shape of each orbital is usually referred to by a letter, rather than by its azimuthal quantum number. The first shape (l=0) is denoted by the letter s (a mnemonic being "sphere"). The next shape is denoted by the letter p and has the form of a dumbbell. The other orbitals have more complicated shapes (see atomic orbital), and are denoted by the letters d, f, g, etc.

The third quantum number, the magnetic quantum number, describes the magnetic moment of the electron, and is denoted by m_l (or simply m). The possible values for m_l are integers from –l to l (where l is the azimuthal quantum number of the electron):

$$m_l = -l, -(l-1), \ldots, 0, 1, \ldots, l.$$

The magnetic quantum number measures the component of the angular momentum in a particular direction. The choice of direction is arbitrary, conventionally the z-direction is chosen.

The fourth quantum number, the spin quantum number (pertaining to the "orientation" of the electron's spin) is denoted m_s, with values $+\frac{1}{2}$ or $-\frac{1}{2}$.

The chemist Linus Pauling wrote, by way of example:

In the case of a helium atom with two electrons in the 1s orbital, the Pauli Exclusion Principle requires that the two electrons differ in the value of one quantum number. Their values of n, l, and m_l are the same. Accordingly they must differ in the value of m_s, which can have the value of $+\frac{1}{2}$ for one electron and $-\frac{1}{2}$ for the other."

It is the underlying structure and symmetry of atomic orbitals, and the way that electrons fill them, that leads to the organisation of the periodic table. The way the atomic orbitals on different atoms combine to form molecular orbitals determines the structure and strength of chemical bonds between atoms.

DIRAC WAVE EQUATION

Paul Dirac (1902–1984)

In 1928, Paul Dirac extended the Pauli equation, which described spinning electrons, to account for special relativity. The result was a theory that dealt properly with events, such as the speed at which an electron orbits the nucleus, occurring at a substantial fraction of the speed of light. By using the simplest electromagnetic interaction, Dirac was able to predict the value of the magnetic moment associated with the electron's spin, and found the experimentally observed value, which was too large to be that of a spinning charged sphere governed by classical physics. He was able to solve for the spectral lines of the hydrogen atom, and to reproduce from physical first principles Sommerfeld's successful formula for the fine structure of the hydrogen spectrum.

Dirac's equations sometimes yielded a negative value for energy, for which he proposed a novel solution: he posited the existence of an antielectron and of a dynamical vacuum. This led to the many-particle quantum field theory.

QUANTUM ENTANGLEMENT

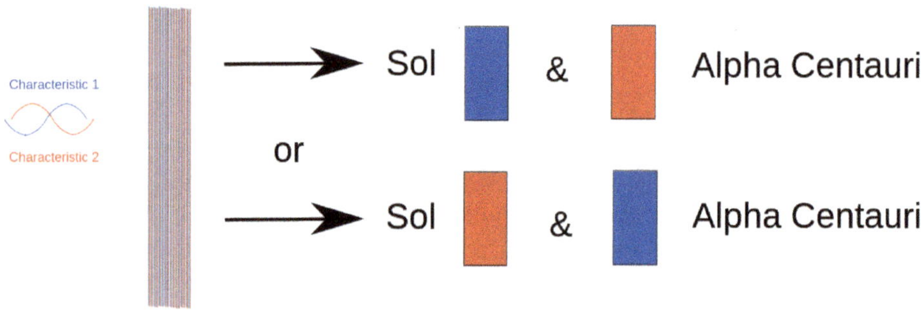

Superposition of two quantum characteristics, and two resolution possibilities.

The Pauli exclusion principle says that two electrons in one system cannot be in the same state. Nature leaves open the possibility, however, that two electrons can have both states "superimposed" over each of them. Recall that the wave functions that emerge simultaneously from the double slits arrive at the detection screen in a state of superposition. Nothing is certain until the superimposed waveforms "collapse". At that instant an electron shows up somewhere in accordance with the probability that is the square of the absolute value of the sum of the complex-valued amplitudes of the two superimposed waveforms. The situation there is already very abstract. A concrete way of thinking about entangled photons, photons in which two contrary states are superimposed on each of them in the same event, is as follows:

Imagine that the superposition of a state that can be mentally **labeled** as blue and another state that can be mentally labeled as red will then appear (in imagination, of course) as a purple state. Two photons are produced as the result of the same atomic event. Perhaps they are produced by the excitation of a crystal that characteristically absorbs a photon of a certain frequency and emits two photons of half the original frequency. So the two photons come out "purple." If the experimenter now performs some experiment that will determine whether one of the photons is either blue or red, then that experiment changes the photon involved from one having a superposition of "blue" and "red" characteristics to a photon that has only one of those characteristics. The problem that Einstein had with such an imagined situation was that if one of these photons had been kept bouncing between mirrors in a laboratory on earth, and the other one had traveled halfway to the nearest star, when its twin was made to reveal itself as either blue or red, that meant that the distant photon now had to lose its purple status too. So whenever it might be investigated after its twin had been measured, it would necessarily show up in the opposite state to whatever its twin had revealed.

In trying to show that quantum mechanics was not a complete theory, Einstein started with the theory's prediction that two or more particles that have interacted in the past can appear strongly correlated when their various properties are later measured. He sought to explain this seeming interaction in a classical way, through their common past, and preferably not by some spooky action at a distance. The argument is worked out in a famous paper, Einstein, Podolsky, and Rosen (1935; abbreviated EPR), setting out what is now called the EPR paradox. Assuming what is now usually called local realism, EPR attempted to show from quantum theory that a particle has both position and momentum simultaneously, while according to the Copenhagen interpretation, only one of those two properties actually exists and only at the moment that it is being measured. EPR concluded that quantum theory is incomplete in that it refuses to consider physical properties which objectively exist in nature. (Einstein, Podolsky, & Rosen 1935 is currently Einstein's most cited publication in physics journals.) In the same year, Erwin Schrödinger used the word entanglement and declared: I would not call that one but rather the characteristic trait of quantum mechanics." The question of whether entanglement is a real condition is still in dispute. The Bell inequalities are the most powerful challenge to Einstein's claims.

QUANTUM FIELD THEORY

The idea of quantum field theory began in the late 1920s with British physicist Paul Dirac, when he attempted to quantise the electromagnetic field – a procedure for constructing a quantum theory starting from a classical theory.

A field in physics is "a region or space in which a given effect (such as magnetism) exists." Other effects that manifest themselves as fields are gravitation and static electricity. In 2008, physicist Richard Hammond wrote that

Sometimes we distinguish between quantum mechanics (QM) and quantum field theory (QFT). QM refers to a system in which the number of particles is fixed, and the fields (such as the electromechanical field) are continuous classical entities. QFT ... goes a step further and allows for the creation and annihilation of particles . . . .

He added, however, that quantum mechanics is often used to refer to "the entire notion of quantum view."

In 1931, Dirac proposed the existence of particles that later became known as antimatter. Dirac shared the Nobel Prize in Physics for 1933 with Schrödinger, "for the discovery of new productive forms of atomic theory."

On its face, quantum field theory allows infinite numbers of particles, and leaves it up to the theory itself to predict how many and with which probabilities or numbers they should exist. When developed further, the theory often contradicts observation, so that its creation and annihilation operators can be empirically tied down. Furthermore, empirical conservation laws like that of mass-energy suggest certain constraints on the mathematical form of the theory, which are mathematically speaking finicky. The latter fact both serves to make quantum field theories difficult to handle, but has also lead to further restrictions on admissible forms of the theory; the complications are mentioned below under the rubrik of renormalization.

Quantum Electrodynamics

Quantum electrodynamics (QED) is the name of the quantum theory of the electromagnetic force. Understanding QED begins with understanding electromagnetism. Electromagnetism can be called "electrodynamics" because it is a dynamic interaction between electrical and magnetic forces. Electromagnetism begins with the electric charge.

Electric charges are the sources of, and create, electric fields. An electric field is a field which exerts a force on any particles that carry electric charges, at any point in space. This includes the electron, proton, and even quarks, among others. As a force is exerted, electric charges move, a current flows and a magnetic field is produced. The changing magnetic field, in turn causes electric current (often moving electrons). The physical description of interacting charged particles, electrical currents, electrical fields, and magnetic fields is called electromagnetism.

In 1928 Paul Dirac produced a relativistic quantum theory of electromagnetism. This was the progenitor to modern quantum electrodynamics, in that it had essential ingredients of the modern theory. However, the problem of unsolvable infinities developed in this relativistic quantum theory. Years later, renormalization largely solved this problem. Initially viewed as a suspect, provisional procedure by some of its originators, renormalization eventually was embraced as an important and self-consistent tool in QED and other fields of physics. Also, in the late 1940s Feynman's diagrams depicted all possible interactions pertaining to a given event. The diagrams showed that the electromagnetic force is the interaction of photons between interacting particles.

An example of a prediction of quantum electrodynamics which has been verified experimentally is the Lamb shift. This refers to an effect whereby the quantum nature of the electromagnetic field causes the energy levels in an atom or ion to deviate slightly from what they would otherwise be. As a result, spectral lines may shift or split.

Similarly, within a freely propagating electromagnetic wave, the current can also be just an abstract displacement current, instead of involving charge carriers. In QED, its full description makes essential use of short lived virtual particles. There, QED again validates an earlier, rather mysterious concept.

Standard Model

In the 1960s physicists realized that QED broke down at extremely high energies. From this inconsistency the Standard Model of particle physics was discovered, which remedied the higher energy breakdown in theory. It is another, extended quantum field theory which unifies the electromagnetic and weak interactions into one theory. This is called the electroweak theory.

Additionally the Standard Model contains a high energy unification of the electroweak theory with the strong force, described by quantum chromodynamics. It also postulates a connection with gravity as yet another gauge theory, but the connection is as of 2015 still poorly understood. The theory's prediction of the Higgs particle to explain inertial mass has stood recent empirical tests at the Large hadron collider, and thus the Standard model is now considered the basic and more or less complete description of particle physics as we know it.

INTERPRETATIONS

The physical measurements, equations, and predictions pertinent to quantum mechanics are all consistent and hold a very high level of confirmation. However, the question of what these abstract models say about the underlying nature of the real world has received competing answers.

Applications

Applications of quantum mechanics include the laser, the transistor, the electron microscope, and magnetic resonance imaging. A special class of quantum mechanical applications is related to macroscopic quantum phenomena such as superfluid helium and superconductors. The study of semiconductors led to the invention of the diode and the transistor, which are indispensable for modern electronics.

In even the simple light switch, quantum tunnelling is absolutely vital, as otherwise the electrons in the electric current could not penetrate the potential barrier made up of a layer of oxide. Flash memory chips found in USB drives also use quantum tunnelling, to erase their memory cells.

QUANTUM MECHANICS

Quantum mechanics (QM; also known as quantum physics or quantum theory), including quantum field theory, is a fundamental branch of physics concerned with processes involving, for example, atoms and photons. Systems such as these which obey quantum mechanics can be in a quantum superposition of different states, unlike in classical physics.

Quantum mechanics gradually arose from Max Planck's solution in 1900 to the black-body radiation problem (reported 1859) and Albert Einstein's 1905 paper which offered a quantum-based theory to explain the photoelectric effect (reported 1887). Early quantum theory was profoundly reconceived in the mid-1920s.

The reconceived theory is formulated in various specially developed mathematical formalisms. In one of them, a mathematical function, the wave function, provides information about the probability amplitude of position, momentum, and other physical properties of a particle.

Important applications of quantum theory include superconducting magnets, light-emitting diodes, and the laser, the transistor and semiconductors such as the microprocessor, medical and research imaging such as magnetic resonance imaging and electron microscopy, and explanations for many biological and physical phenomena.[1]

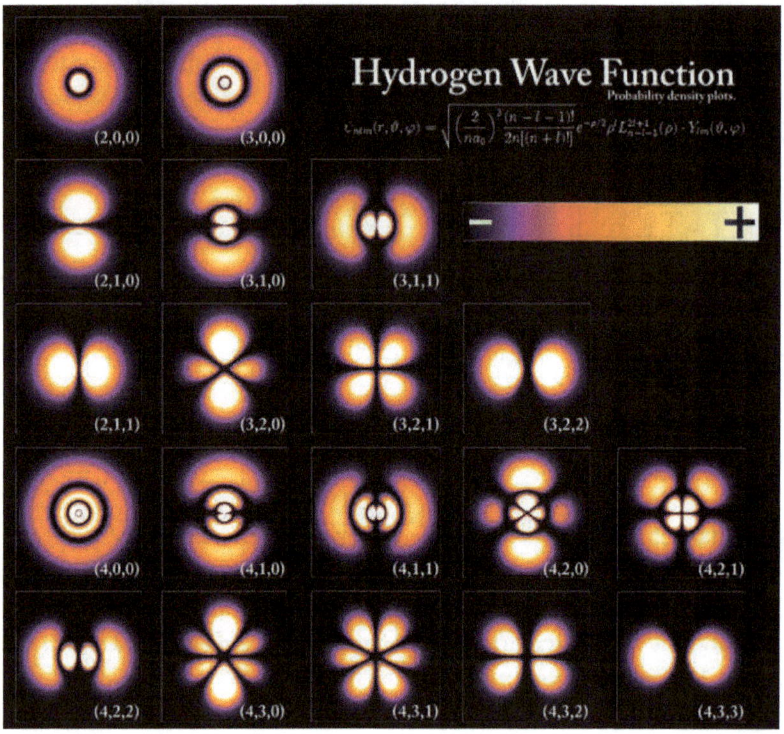

Solution to Schrödinger's equation for the hydrogen atom at different energy levels. The brighter areas represent a higher probability of finding an electron.

Scientific inquiry into the wave nature of light began in the 17th and 18th centuries, when scientists such as Robert Hooke, Christiaan Huygens and Leonhard Euler proposed a wave theory of light based on experimental observations. In 1803, Thomas Young, an English polymath, performed the famous double-slit experiment that he later described in a paper titled *On the nature of light and colours*. This experiment played a major role in the general acceptance of the wave theory of light.

In 1838, Michael Faraday discovered cathode rays. These studies were followed by the 1859 statement of the black-body radiation problem by Gustav Kirchhoff, the 1877 suggestion by Ludwig Boltzmann that the energy states of a physical system can be discrete, and the 1900 quantum hypothesis of Max Planck. Planck's hypothesis that energy is radiated and absorbed in discrete "quanta" (or energy packets) precisely matched the observed patterns of black-body radiation.

In 1896, Wilhelm Wien empirically determined a distribution law of black-body radiation, known as Wien's law in his honor. Ludwig Boltzmann independently arrived at this result by considerations of Maxwell's equations. However, it was valid only at high frequencies and underestimated the radiance at low frequencies. Later, Planck corrected this model using Boltzmann's statistical interpretation of thermodynamics and proposed what is now called Planck's law, which led to the development of quantum mechanics.

Following Max Planck's solution in 1900 to the black-body radiation problem (reported 1859), Albert Einstein offered a quantum-based theory to explain the photoelectric effect (1905, reported 1887). Around 1900-1910, the atomic theory and the corpuscular theory of light first came to be widely accepted as scientific fact; these latter theories can be viewed as quantum theories of matter and electromagnetic radiation, respectively.

Among the first to study quantum phenomena in nature were Arthur Compton, C. V. Raman, and Pieter Zeeman, each of whom has a quantum effect named after him. Robert Andrews Millikan studied the photoelectric effect experimentally, and Albert Einstein developed a theory for it. At the same time, Ernest Rutherford experimentally discovered the nuclear model of the atom, for which Niels Bohr developed his theory of the atomic structure, which was later confirmed by the experiments of Henry Moseley. In 1913, Peter Debye extended Niels Bohr's theory of atomic structure, introducing elliptical orbits, a concept also introduced by Arnold Sommerfeld. This phase is known as old quantum theory.

According to Planck, each energy element (E) is proportional to its frequency (ν):

$E = h\nu$ where h is Planck's constant.

Max Planck is considered the father of the quantum theory.

Planck cautiously insisted that this was simply an aspect of the *processes* of absorption and emission of radiation and had nothing to do with the *physical reality* of the radiation itself. In fact, he considered his quantum hypothesis a mathematical trick to get the right answer rather than a sizable discovery. However, in 1905 Albert Einstein interpreted Planck's quantum hypothesis realistically and used it to explain the photoelectric effect, in which shining light on certain materials can eject electrons from the material. He won the 1921 Nobel Prize in Physics for this work.

Einstein further developed this idea to show that an electromagnetic wave such as light could also be described as a particle (later called the photon), with a discrete quantum of energy that was dependent on its frequency.

The 1927 Solvay Conference in Brussels.

The foundations of quantum mechanics were established during the first half of the 20th century by Max Planck, Niels Bohr, Werner Heisenberg, Louis de Broglie, Arthur Compton, Albert Einstein, Erwin Schrödinger, Max Born, John von Neumann, Paul Dirac, Enrico Fermi, Wolfgang Pauli, Max von Laue, Freeman Dyson, David Hilbert, Wilhelm Wien, Satyendra Nath Bose, Arnold Sommerfeld, and others. The Copenhagen interpretation of Niels Bohr became widely accepted.

In the mid-1920s, developments in quantum mechanics led to its becoming the standard formulation for atomic physics. In the summer of 1925, Bohr and Heisenberg published results that closed the old quantum theory. Out of deference to their particle-like behavior in certain processes and measurements, light quanta came to be called photons (1926). From Einstein's simple postulation was born a flurry of debating, theorizing, and testing. Thus, the entire field of quantum physics emerged, leading to its wider acceptance at the Fifth Solvay Conference in 1927.

It was found that subatomic particles and electromagnetic waves are neither simply particle nor wave but have certain properties of each. This originated the concept of wave–particle duality.

By 1930, quantum mechanics had been further unified and formalized by the work of David Hilbert, Paul Dirac and John von Neumann with greater emphasis on measurement, the statistical nature of our knowledge of reality, and philosophical speculation about the 'observer'. It has since permeated many disciplines including quantum chemistry, quantum electronics, quantum optics, and quantum information science. Its speculative modern developments include string theory and quantum gravity theories. It also provides a useful framework for many features of the modern periodic table of elements, and describes the behaviors of atoms during chemical bonding and the flow of electrons in computer semiconductors, and therefore plays a crucial role in many modern technologies.

While quantum mechanics was constructed to describe the world of the very small, it is also needed to explain some macroscopic phenomena such as superconductors, and superfluids.

The word *quantum* derives from the Latin, meaning "how great" or "how much". In quantum mechanics, it refers to a discrete unit assigned to certain physical quantities such as the energy of an atom at rest (see Figure 1). The discovery that particles are discrete packets of energy with wave-like properties led to the branch of physics dealing with atomic and subatomic systems which is today called quantum mechanics. It underlies the mathematical framework of many fields of physics and chemistry, including condensed matter physics, solid-state physics, atomic physics, molecular physics, computational physics, computational chemistry, quantum chemistry, particle physics, nuclear chemistry, and nuclear physics. Some fundamental aspects of the theory are still actively studied.

Quantum mechanics is essential to understanding the behavior of systems at atomic length scales and smaller. If the physical nature of an atom were solely described by classical mechanics, electrons would not *orbit* the nucleus, since orbiting electrons emit radiation (due to circular motion) and would eventually collide with the nucleus due to this loss of energy. This framework was unable to explain the stability of atoms. Instead, electrons remain in an uncertain, non-deterministic, *smeared*, probabilistic wave–particle orbital about the nucleus, defying the traditional assumptions of classical mechanics and electromagnetism.

Quantum mechanics was initially developed to provide a better explanation and description of the atom, especially the differences in the spectra of light emitted by different isotopes of the same chemical element, as well as subatomic particles. In short, the quantum-mechanical atomic model has succeeded spectacularly in the realm where classical mechanics and electromagnetism falter.

Broadly speaking, quantum mechanics incorporates four classes of phenomena for which classical physics cannot account:

- quantization of certain physical properties
- quantum entanglement
- principle of uncertainty
- wave–particle duality

MATHEMATICAL FORMULATIONS

In the mathematically rigorous formulation of quantum mechanics developed by Paul Dirac, David Hilbert, John von Neumann, and Hermann Weyl, the possible states of a quantum mechanical system are symbolized as unit vectors (called *state vectors*). Formally, these reside in a complex separable Hilbert space—variously called the *state space* or the *associated Hilbert space* of the system—that is well defined up to a complex number of norm 1 (the phase factor). In other words, the possible states are points in the projective space of a Hilbert space, usually called the complex projective space. The exact nature of this Hilbert space is dependent on the system—for example, the state space for position and momentum states is the space of square-integrable functions, while the state space for the spin of a single proton is just the product of two complex planes. Each observable is represented by a maximally Hermitian (precisely: by a self-adjoint) linear operator acting on the state space. Each eigenstate of an observable corresponds to an eigenvector of the operator, and the associated eigenvalue corresponds to the value of the observable in that eigenstate. If the operator's spectrum is discrete, the observable can attain only those discrete eigenvalues.

In the formalism of quantum mechanics, the state of a system at a given time is described by a complex wave function, also referred to as state vector in a complex vector space. This abstract mathematical object allows for the calculation of probabilities of outcomes of concrete experiments. For example, it allows one to compute the probability of finding an electron in a particular region around the nucleus at a particular time. Contrary to classical mechanics, one can never make simultaneous predictions of conjugate variables, such as position and momentum, to arbitrary precision. For instance, electrons may be considered (to a certain probability) to be located somewhere within a given region of space, but with their exact positions unknown. Contours of constant probability, often referred to as clouds, may be drawn around the nucleus of an atom to conceptualize where the electron might be located with the most probability. Heisenberg's uncertainty principle quantifies the inability to precisely locate the particle given its conjugate momentum.

According to one interpretation, as the result of a measurement the wave function containing the probability information for a system collapses from a given initial state to a particular eigenstate. The possible results of a measurement are the eigenvalues of the operator representing the observable—which explains the choice of *Hermitian* operators, for which all the eigenvalues are real. The probability distribution of an observable in a given state can be found by computing the spectral decomposition of the corresponding operator. Heisenberg's uncertainty principle is represented by the statement that the operators corresponding to certain observables do not commute.

The probabilistic nature of quantum mechanics thus stems from the act of measurement. This is one of the most difficult aspects of quantum systems to understand. It was the central topic in the famous Bohr–Einstein debates, in which the two scientists attempted to clarify these fundamental principles by way of thought experiments. In the decades after the formulation of quantum mechanics, the question of what constitutes a measurement has been extensively studied. Newer interpretations of quantum mechanics have been formulated that do away with the concept of wave function collapse (see, for example, the relative state interpretation). The basic idea is that when a quantum system interacts with a measuring apparatus, their respective wave functions become entangled, so that the original quantum system ceases to exist as an independent entity. For details, see the article on measurement in quantum mechanics.

Generally, quantum mechanics does not assign definite values. Instead, it makes a prediction using a probability distribution; that is, it describes the probability of obtaining the possible outcomes from measuring an observable. Often these results are skewed by many causes, such as dense probability clouds. Probability clouds are approximate (but better than the Bohr model) whereby electron location is given by a probability function, the wave function eigenvalue, such that the probability is the squared modulus of the complex amplitude, or quantum state nuclear attraction.Naturally, these probabilities will depend on the quantum state at the instant of the measurement. Hence, uncertainty is involved in the value. There are, however, certain states that are associated with a definite value of a particular observable. These are known as eigenstates of the observable (eigen can be translated from German as meaning inherent or characteristic).

In the everyday world, it is natural and intuitive to think of everything (every observable) as being in an eigenstate. Everything appears to have a definite position, a definite momentum, a definite energy, and a definite time of occurrence. However, quantum mechanics does not pinpoint the exact values of a particle's position and momentum (since they are conjugate pairs) or its energy and time (since they too are conjugate pairs); rather, it provides only a range of probabilities in which that particle might be given its momentum and momentum probability. Therefore, it is helpful to use different words to describe states having *uncertain* values and states having *definite* values (eigenstates). Usually, a system will not be in an eigenstate of the observable (particle) we are interested in. However, if one measures the observable, the wave function will instantaneously be an eigenstate (or "generalized" eigenstate) of that observable. This process is known as wave function collapse, a controversial and much-debated process that involves expanding the system under study to include the measurement device. If one knows the corresponding wave function at the instant before the measurement, one will be able to compute the probability of the wave function collapsing into each of the possible eigenstates. For example, the free particle in the previous example will usually have a wave function that is a wave packet centered around some mean position x_0 (neither an eigenstate of position nor of momentum). When one measures the position of the particle, it is impossible to predict with certainty the result. It is probable, but not certain, that it will be near x_0, where the amplitude of the wave function is large. After the measurement is performed, having obtained some result x, the wave function collapses into a position eigenstate centered at x.

The time evolution of a quantum state is described by the Schrödinger equation, in which the Hamiltonian (the operator corresponding to the total energy of the system) generates the time evolution. The time evolution of wave functions is deterministic in the sense that - given a wave function at an *initial* time - it makes a definite prediction of what the wave function will be at any *later* time.

During a measurement, on the other hand, the change of the initial wave function into another, later wave function is not deterministic, it is unpredictable (i.e., random). A time-evolution simulation can be seen here.

Wave functions change as time progresses. The Schrödinger equation describes how wave functions change in time, playing a role similar to Newton's second law in classical mechanics. The Schrödinger equation, applied to the aforementioned example of the free particle, predicts that the center of a wave packet will move through space at a constant velocity (like a classical particle with no forces acting on it). However, the wave packet will also spread out as time progresses, which means that the position becomes more uncertain with time. This also has the effect of turning a position eigenstate (which can be thought of as an infinitely sharp wave packet) into a broadened wave packet that no longer represents a (definite, certain) position eigenstate.

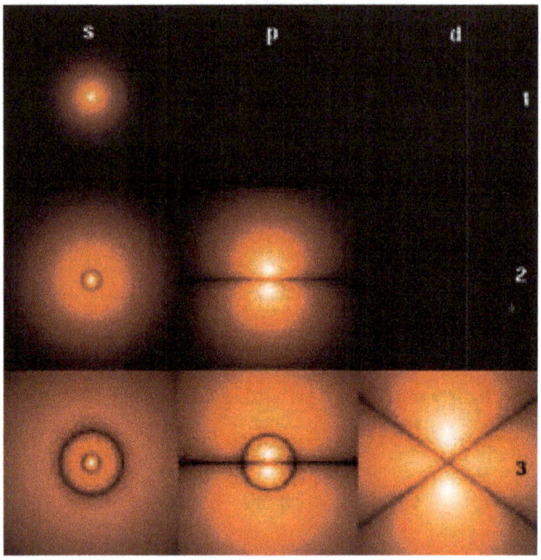

Fig. 1: Probability densities corresponding to the wave functions of an electron in a hydrogen atom possessing definite energy levels (increasing from the top of the image to the bottom: $n = 1, 2, 3, \ldots$) and angular momenta (increasing across from left to right: s, p, d, ...). Brighter areas correspond to higher probability density in a position measurement. Such wave functions are directly comparable to Chladni's figures of acoustic modes of vibration in classical physics, and are modes of oscillation as well, possessing a sharp energy and, thus, a definite frequency. The angular momentum and energy are quantized, and take **only** discrete values like those shown (as is the case for resonant frequencies in acoustics)

Some wave functions produce probability distributions that are constant, or independent of time—such as when in a stationary state of constant energy, time vanishes in the absolute square of the wave function. Many systems that are treated dynamically in classical mechanics are described by such static wave functions. For example, a single electron in an unexcited atom is pictured classically as a particle moving in a circular trajectory around the atomic nucleus, whereas in quantum mechanics it is described by a static, spherically symmetric wave function surrounding the nucleus (Fig. 1) (note, however, that only the lowest angular momentum states, labeled s, are spherically symmetric).

The Schrödinger equation acts on the *entire* probability amplitude, not merely its absolute value. Whereas the absolute value of the probability amplitude encodes information about probabilities, its phase encodes information about the interference between quantum states. This gives rise to the “wave-like” behavior of quantum states. As it turns out, analytic solutions of the Schrödinger equation are available for only a very small number of relatively simple model Hamiltonians, of which the quantum harmonic oscillator, the particle in a box, the dihydrogen cation, and the hydrogen atom are the most important representatives. Even the helium atom—which contains just one more electron than does the hydrogen atom—has defied all attempts at a fully analytic treatment.

There exist several techniques for generating approximate solutions, however. In the important method known as perturbation theory, one uses the analytic result for a simple quantum mechanical model to generate a result for a more complicated model that is related to the simpler model by (for one example) the addition of a weak potential energy. Another method is the semi-classical equation of motion approach, which applies to systems for which quantum mechanics produces only weak (small) deviations from classical behavior. These deviations can then be computed based on the classical motion. This approach is particularly important in the field of quantum chaos.

MATHEMATICALLY EQUIVALENT FORMULATIONS OF QUANTUM MECHANICS

There are numerous mathematically equivalent formulations of quantum mechanics. One of the oldest and most commonly used formulations is the “transformation theory” proposed by Paul Dirac, which unifies and generalizes the two earliest formulations of quantum mechanics - matrix mechanics (invented by Werner Heisenberg) and wave mechanics (invented by Erwin Schrödinger).

Especially since Werner Heisenberg was awarded the Nobel Prize in Physics in 1932 for the creation of quantum mechanics, the role of Max Born in the development of QM was overlooked until the 1954 Nobel award. The role is noted in a 2005 biography of Born, which recounts his role in the matrix formulation of quantum mechanics, and the use of probability amplitudes. Heisenberg himself acknowledges having learned matrices from Born, as published in a 1940 *festschrift* honoring Max Planck. In the matrix formulation, the instantaneous state of a quantum system encodes the probabilities of its measurable properties, or “observables”. Examples of observables include energy, position, momentum, and angular momentum. Observables can be either continuous (e.g., the position of a particle) or discrete (e.g., the energy of an electron bound to a hydrogen atom). An alternative formulation of quantum mechanics is Feynman’s path integral formulation, in which a quantum-mechanical amplitude is considered as a sum over all possible classical and non-classical paths between the initial and final states. This is the quantum-mechanical counterpart of the action principle in classical mechanics.

INTERACTIONS WITH OTHER SCIENTIFIC THEORIES

The rules of quantum mechanics are fundamental. They assert that the state space of a system is a Hilbert space and that observables of that system are Hermitian operators acting on that space—although they do not tell us which Hilbert space or which operators. These can be chosen appropriately in order to obtain a quantitative description of a quantum system. An important guide for making these choices is the correspondence principle, which states that the predictions of quantum mechanics reduce to those of classical mechanics when a system moves to higher energies or, equivalently, larger quantum numbers, i.e. whereas a single particle exhibits a degree of randomness, in systems incorporating millions of particles averaging takes over and, at the high energy limit, the statistical probability of random behaviour approaches zero. In other words, classical mechanics is simply a quantum mechanics of large systems. This "high energy" limit is known as the *classical* or *correspondence limit*. One can even start from an established classical model of a particular system, then attempt to guess the underlying quantum model that would give rise to the classical model in the correspondence limit.

When quantum mechanics was originally formulated, it was applied to models whose correspondence limit was non-relativistic classical mechanics. For instance, the well-known model of the quantum harmonic oscillator uses an explicitly non-relativistic expression for the kinetic energy of the oscillator, and is thus a quantum version of the classical harmonic oscillator.

Early attempts to merge quantum mechanics with special relativity involved the replacement of the Schrödinger equation with a covariant equation such as the Klein–Gordon equation or the Dirac equation. While these theories were successful in explaining many experimental results, they had certain unsatisfactory qualities stemming from their neglect of the relativistic creation and annihilation of particles. A fully relativistic quantum theory required the development of quantum field theory, which applies quantization to a field (rather than a fixed set of particles). The first complete quantum field theory, quantum electrodynamics, provides a fully quantum description of the electromagnetic interaction. The full apparatus of quantum field theory is often unnecessary for describing electrodynamic systems. A simpler approach, one that has been employed since the inception of quantum mechanics, is to treat charged particles as quantum mechanical objects being acted on by a classical electromagnetic field. For example, the elementary quantum model of the hydrogen atom describes the electric field of the hydrogen atom using a classical $-e^2/(4\pi \epsilon_0 r)$ Coulomb potential. This "semi-classical" approach fails if quantum fluctuations in the electromagnetic field play an important role, such as in the emission of photons by charged particles.

Quantum field theories for the strong nuclear force and the weak nuclear force have also been developed. The quantum field theory of the strong nuclear force is called quantum chromodynamics, and describes the interactions of subnuclear particles such as quarks and gluons. The weak nuclear force and the electromagnetic force were unified, in their quantized forms, into a single quantum field theory (known as electroweak theory), by the physicists Abdus Salam, Sheldon Glashow and Steven Weinberg. These three men shared the Nobel Prize in Physics in 1979 for this work.

It has proven difficult to construct quantum models of gravity, the remaining fundamental force. Semi-classical approximations are workable, and have led to predictions such as Hawking radiation. However, the formulation of a complete theory of quantum gravity is hindered by apparent incompatibilities between general relativity (the most accurate theory of gravity currently known) and some of the fundamental assumptions of quantum theory. The resolution of these incompatibilities is an area of active research, and theories such as string theory are among the possible candidates for a future theory of quantum gravity.

Classical mechanics has also been extended into the complex domain, with complex classical mechanics exhibiting behaviors similar to quantum mechanics.

Quantum Mechanics and Classical Physics

Predictions of quantum mechanics have been verified experimentally to an extremely high degree of accuracy. According to the correspondence principle between classical and quantum mechanics, all objects obey the laws of quantum mechanics, and classical mechanics is just an approximation for large systems of objects (or a statistical quantum mechanics of a large collection of particles). The laws of classical mechanics thus follow from the laws of quantum mechanics as a statistical average at the limit of large systems or large quantum numbers.However, chaotic systems do not have good quantum numbers, and quantum chaos studies the relationship between classical and quantum descriptions in these systems.

Quantum coherence is an essential difference between classical and quantum theories as illustrated by the Einstein–Podolsky–Rosen (EPR) paradox — an attack on a certain philosophical interpretation of quantum mechanics by an appeal to local realism. Quantum interference involves adding together *probability amplitudes*, whereas classical "waves" infer that there is an adding together of *intensities*. For microscopic bodies, the extension of the system is much smaller than the coherence length, which gives rise to long-range entanglement and other nonlocal phenomena characteristic of quantum systems. Quantum coherence is not typically evident at macroscopic scales, though an exception to this rule may occur at extremely low temperatures (i.e. approaching absolute zero) at which quantum behavior may manifest itself macroscopically.This is in accordance with the following observations:

- Many macroscopic properties of a classical system are a direct consequence of the quantum behavior of its parts. For example, the stability of bulk matter (consisting of atoms and molecules which would quickly collapse under electric forces alone), the rigidity of solids, and the mechanical, thermal, chemical, optical and magnetic properties of matter are all results of the interaction of electric charges under the rules of quantum mechanics.
- While the seemingly "exotic" behavior of matter posited by quantum mechanics and relativity theory become more apparent when dealing with particles of extremely small size or velocities approaching the speed of light, the laws of classical, often considered "Newtonian", physics remain accurate in predicting the behavior of the vast majority of "large" objects (on the order of the size of large molecules or bigger) at velocities much smaller than the velocity of light.

Copenhagen Interpretation of Quantum Versus Classical Kinematics

A big difference between classical and quantum mechanics is that they use very different kinematic descriptions.

In Niels Bohr's mature view, quantum mechanical phenomena are required to be experiments, with complete descriptions of all the devices for the system, preparative, intermediary, and finally measuring. The descriptions are in macroscopic terms, expressed in ordinary language, supplemented with the concepts of classical mechanics. The initial condition and the final condition of the system are respectively described by values in a configuration space, for example a position space, or some equivalent space such as a momentum space. Quantum mechanics does not admit a completely precise description, in terms of both position and momentum, of an initial condition or "state" (in the classical sense of the word) that would support a precisely deterministic and causal prediction of a final condition. In this sense, advocated by Bohr in his mature writings, a quantum phenomenon is a process, a passage from initial to final condition, not an instantaneous "state" in the classical sense of that word. Thus there are two kinds of processes in quantum mechanics: stationary and transitional. For a stationary process, the initial and final condition are the same. For a transition, they are different. Obviously by definition, if only the initial condition is given, the process is not determined. Given its initial condition, prediction of its final condition is possible, causally but only probabilistically, because the Schrödinger equation is deterministic for wave function evolution, but the wave function describes the system only probabilistically.

For many experiments, it is possible to think of the initial and final conditions of the system as being a particle. In some cases it appears that there are potentially several spatially distinct pathways or trajectories by which a particle might pass from initial to final condition. It is an important feature of the quantum kinematic description that it does not permit a unique definite statement of which of those pathways is actually followed. Only the initial and final conditions are definite, and, as stated in the foregoing paragraph, they are defined only as precisely as allowed by the configuration space description or its equivalent. In every case for which a quantum kinematic description is needed, there is always a compelling reason for this restriction of kinematic precision. An example of such a reason is that for a particle to be experimentally found in a definite position, it must be held motionless; for it to be experimentally found to have a definite momentum, it must have free motion; these two are logically incompatible.

Classical kinematics does not primarily demand experimental description of its phenomena. It allows completely precise description of an instantaneous state by a value in phase space, the Cartesian product of configuration and momentum spaces. This description simply assumes or imagines a state as a physically existing entity without concern about its experimental measurability. Such a description of an initial condition, together with Newton's laws of motion, allows a precise deterministic and causal prediction of a final condition, with a definite trajectory of passage. Hamiltonian dynamics can be used for this. Classical kinematics also allows the description of a process analogous to the initial and final condition description used by quantum mechanics. Lagrangian mechanics applies to this. For processes that need account to be taken of actions of a small number of Planck constants, classical kinematics is not adequate; quantum mechanics is needed.

General Relativity and Quantum Mechanics

Even with the defining postulates of both Einstein's theory of general relativity and quantum theory being indisputably supported by rigorous and repeated empirical evidence, and while they do not directly contradict each other theoretically (at least with regard to their primary claims), they have proven extremely difficult to incorporate into one consistent, cohesive model.

Gravity is negligible in many areas of particle physics, so that unification between general relativity and quantum mechanics is not an urgent issue in those particular applications. However, the lack of a correct theory of quantum gravity is an important issue in physical cosmology and the search by physicists for an elegant "Theory of Everything" (TOE). Consequently, resolving the inconsistencies between both theories has been a major goal of 20th and 21st century physics. Many prominent physicists, including Stephen Hawking, have labored for many years in the attempt to discover a theory underlying *everything*. This TOE would combine not only the different models of subatomic physics, but also derive the four fundamental forces of nature - the strong force, electromagnetism, the weak force, and gravity - from a single force or phenomenon. While Stephen Hawking was initially a believer in the Theory of Everything, after considering Gödel's Incompleteness Theorem, he has concluded that one is not obtainable, and has stated so publicly in his lecture "Gödel and the End of Physics" (2002).

Attempts at a Unified Field Theory

The quest to unify the fundamental forces through quantum mechanics is still ongoing. Quantum electrodynamics (or "quantum electromagnetism"), which is currently (in the perturbative regime at least) the most accurately tested physical theory in competition with general relativity, has been successfully merged with the weak nuclear force into the electroweak force and work is currently being done to merge the electroweak and strong force into the electrostrong force. Current predictions state that at around 10^{14} GeV the three aforementioned forces are fused into a single unified field. Beyond this "grand unification", it is speculated that it may be possible to merge gravity with the other three gauge symmetries, expected to occur at roughly 10^{19} GeV. However — and while special relativity is parsimoniously incorporated into quantum electrodynamics — the expanded general relativity, currently the best theory describing the gravitation force, has not been fully incorporated into quantum theory. One of those searching for a coherent TOE is Edward Witten, a theoretical physicist who formulated the M-theory, which is an attempt at describing the supersymmetrical based string theory. M-theory posits that our apparent 4-dimensional spacetime is, in reality, actually an 11-dimensional spacetime containing 10 spatial dimensions and 1 time dimension, although 7 of the spatial dimensions are - at lower energies - completely compactified (or infinitely curved) and not readily amenable to measurement or probing.

Another popular theory is Loop quantum gravity (LQG), a theory first proposed by Carlo Rovelli that describes the quantum properties of gravity. It is also a theory of quantum space and quantum time, because in general relativity the geometry of spacetime is a manifestation of gravity. LQG is an attempt to merge and adapt standard quantum mechanics and standard general relativity. The main output of the theory is a physical picture of space where space is granular. The granularity is a direct consequence of the quantization. It has the same nature of the granularity of the photons in the quantum theory of electromagnetism or the discrete levels of the energy of the atoms. But here it is space itself which is discrete. More precisely, space can be viewed as an extremely fine fabric or network woven of finite loops. These networks of loops are called spin networks. The evolution of a spin network over time is called a spin foam. The predicted size of this structure is the Planck length, which is approximately 1.616×10^{-35} m. According to theory, there is no meaning to length shorter than this (cf. Planck scale energy). Therefore, LQG predicts that not just matter, but also space itself, has an atomic structure.

PHILOSOPHICAL IMPLICATIONS

Since its inception, the many counter-intuitive aspects and results of quantum mechanics have provoked strong philosophical debates and many interpretations. Even fundamental issues, such as Max Born's basic rules concerning probability amplitudes and probability distributions, took decades to be appreciated by society and many leading scientists. Richard Feynman once said, "I think I can safely say that nobody understands quantum mechanics." According to Steven Weinberg, "There is now in my opinion no entirely satisfactory interpretation of quantum mechanics."

The Copenhagen interpretation — due largely to Niels Bohr and Werner Heisenberg — remains most widely accepted amongst physicists, some 75 years after its enunciation. According to this interpretation, the probabilistic nature of quantum mechanics is not a *temporary* feature which will eventually be replaced by a deterministic theory, but instead must be considered a *final* renunciation of the classical idea of "causality." It is also believed therein that any well-defined application of the quantum mechanical formalism must always make reference to the experimental arrangement, due to the conjugate nature of evidence obtained under different experimental situations.

Albert Einstein, himself one of the founders of quantum theory, did not accept some of the more philosophical or metaphysical interpretations of quantum mechanics, such as rejection of determinism and of causality. He is famously quoted as saying, in response to this aspect, God does not play with dice. He rejected the concept that the state of a physical system depends on the experimental arrangement for its measurement. He held that a state of nature occurs in its own right, regardless of whether or how it might be observed. In that view, he is supported by the currently accepted definition of a quantum state, which remains invariant under arbitrary choice of configuration space for its representation, that is to say, manner of observation. He also held that underlying quantum mechanics there should be a theory that thoroughly and directly expresses the rule against action at a distance; in other words, he insisted on the principle of locality. He considered, but rejected on theoretical grounds, a particular proposal for hidden variables to obviate the indeterminism or acausality of quantum mechanical measurement. He considered that quantum mechanics was a currently valid but not a permanently definitive theory for quantum phenomena. He thought its future replacement would require profound conceptual advances, and would not come quickly or easily. The Bohr-Einstein debates provide a vibrant critique of the Copenhagen Interpretation from an epistemological point of view. In arguing for his views, he produced a series of objections, the most famous of which has become known as the Einstein–Podolsky–Rosen paradox.

John Bell showed that this EPR paradox led to experimentally testable differences between quantum mechanics and theories that rely on added hidden variables. Experiments have been performed confirming the accuracy of quantum mechanics, thereby demonstrating that quantum mechanics cannot be improved upon by addition of hidden variables. Alain Aspect's initial experiments in 1982, and many subsequent experiments since, have definitively verified quantum entanglement.

Entanglement, as demonstrated in Bell-type experiments, does not, however, violate causality, since no transfer of information happens. Quantum entanglement forms the basis of quantum cryptography, which is proposed for use in high-security commercial applications in banking and government.

The Everett many-worlds interpretation, formulated in 1956, holds that *all* the possibilities described by quantum theory *simultaneously* occur in a multiverse composed of mostly independent parallel universes. This is not accomplished by introducing some "new axiom" to quantum mechanics, but on the contrary, by *removing* the axiom of the collapse of the wave packet. *All* of the possible consistent states of the measured system and the measuring apparatus (including the observer) are present in a *real* physical - not just formally mathematical, as in other interpretations - quantum superposition. Such a superposition of consistent state combinations of different systems is called an entangled state. While the multiverse is deterministic, we perceive non-deterministic behavior governed by probabilities, because we can only observe the universe (i.e., the consistent state contribution to the aforementioned superposition) that we, as observers, inhabit. Everett's interpretation is perfectly consistent with John Bell's experiments and makes them intuitively understandable. However, according to the theory of quantum decoherence, these parallel universes will never be accessible to us. The inaccessibility can be understood as follows: once a measurement is done, the measured system becomes entangled with *both* the physicist who measured it *and* a huge number of other particles, some of which are photons flying away at the speed of light towards the other end of the universe. In order to prove that the wave function did not collapse, one would have to bring *all* these particles back and measure them again, together with the system that was originally measured. Not only is this completely impractical, but even if one *could* theoretically do this, it would have to destroy any evidence that the original measurement took place (including the physicist's memory). In light of these Bell tests, Cramer (1986) formulated his transactional interpretation. Relational quantum mechanics appeared in the late 1990s as the modern derivative of the Copenhagen Interpretation.

APPLICATIONS

Quantum mechanics has had enormous success in explaining many of the features of our universe. Quantum mechanics is often the only tool available that can reveal the individual behaviors of the subatomic particles that make up all forms of matter (electrons, protons, neutrons, photons, and others). Quantum mechanics has strongly influenced string theories, candidates for a Theory of Everything (see reductionism).

Quantum mechanics is also critically important for understanding how individual atoms combine covalently to form molecules. The application of quantum mechanics to chemistry is known as quantum chemistry. Relativistic quantum mechanics can, in principle, mathematically describe most of chemistry. Quantum mechanics can also provide quantitative insight into ionic and covalent bonding processes by explicitly showing which molecules are energetically favorable to which others and the magnitudes of the energies involved. Furthermore, most of the calculations performed in modern computational chemistry rely on quantum mechanics.

In many aspects modern technology operates at a scale where quantum effects are significant.

Electronics

Many modern electronic devices are designed using quantum mechanics. Examples include the laser, the transistor (and thus the microchip), the electron microscope, and magnetic resonance imaging (MRI). The study of semiconductors led to the invention of the diode and the transistor, which are indispensable parts of modern electronics systems, computer and telecommunication devices. Another application is the light emitting diode which is a high-efficiency source of light.

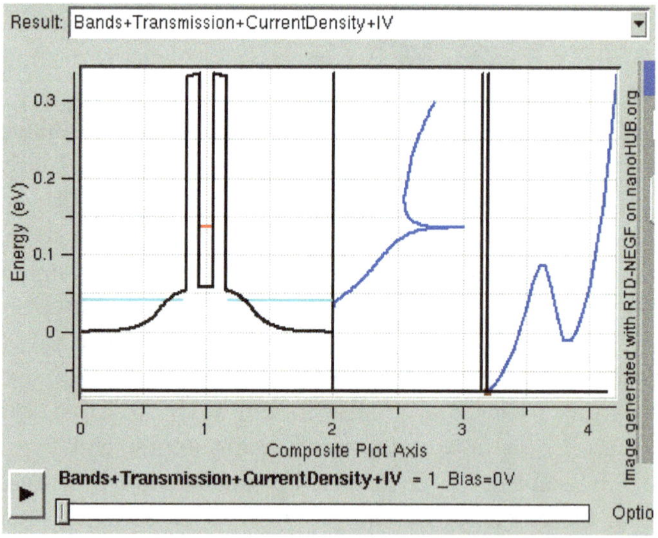

A working mechanism of a resonant tunneling diode device, based on the phenomenon of quantum tunneling through potential barriers. (Left: band diagram; Center: transmission coefficient; Right: current-voltage characteristics) As shown in the band diagram(left), although there are two barriers, electrons still tunnel through via the confined states between two barriers(center), conducting current.

Many electronic devices operate under effect of Quantum tunneling. It even exists in the simple light switch. The switch would not work if electrons could not quantum tunnel through the layer of oxidation on the metal contact surfaces. Flash memory chips found in USB drives use quantum tunneling to erase their memory cells. Some negative differential resistance devices also utilizes quantum tunneling effect, such as resonant tunneling diode. Unlike classical diodes, its current is carried by resonant tunneling through two potential barriers (see right figure). Its negative resistance behavior can only be understood with quantum mechanics: As the confined state moves close to Fermi level, tunnel current increases. As it moves away, current decreases. Quantum mechanics is vital to understanding and designing such electronic devices.

Cryptography

Researchers are currently seeking robust methods of directly manipulating quantum states. Efforts are being made to more fully develop quantum cryptography, which will theoretically allow guaranteed secure transmission of information.

Quantum Computing

A more distant goal is the development of quantum computers, which are expected to perform certain computational tasks exponentially faster than classical computers. Instead of using classical bits, quantum computers use qubits, which can be in superpositions of states. Another active research topic is quantum teleportation, which deals with techniques to transmit quantum information over arbitrary distances.

Macroscale Quantum Effects

While quantum mechanics primarily applies to the smaller atomic regimes of matter and energy, some systems exhibit quantum mechanical effects on a large scale. Superfluidity, the frictionless flow of a liquid at temperatures near absolute zero, is one well-known example. So is the closely related phenomenon of superconductivity, the frictionless flow of an electron gas in a conducting material (an electric current) at sufficiently low temperatures. The fractional quantum hall effect is a topological ordered state which corresponds to patterns of long-range quantum entanglement. States with different topological orders (or different patterns of long range entanglements) cannot change into each other without a phase transition.

Quantum Theory

Quantum theory also provides accurate descriptions for many previously unexplained phenomena, such as black-body radiation and the stability of the orbitals of electrons in atoms. It has also given insight into the workings of many different biological systems, including smell receptors and protein structures. Recent work on photosynthesis has provided evidence that quantum correlations play an essential role in this fundamental process of plants and many other organisms. Even so, classical physics can often provide good approximations to results otherwise obtained by quantum physics, typically in circumstances with large numbers of particles or large quantum numbers. Since classical formulas are much simpler and easier to compute than quantum formulas, classical approximations are used and preferred when the system is large enough to render the effects of quantum mechanics insignificant.

Examples

Free Particle

For example, consider a free particle. In quantum mechanics, there is wave–particle duality, so the properties of the particle can be described as the properties of a wave. Therefore, its quantum state can be represented as a wave of arbitrary shape and extending over space as a wave function. The position and momentum of the particle are observables. The Uncertainty Principle states that both the position and the momentum cannot simultaneously be measured with complete precision. However, one *can* measure the position (alone) of a moving free particle, creating an eigenstate of position with a wave function that is very large (a Dirac delta) at a particular position x, and zero everywhere else. If one performs a position measurement on such a wave function, the resultant x will be obtained with 100% probability (i.e., with full certainty, or complete precision). This is called an eigenstate of position—or, stated in mathematical terms, a *generalized position eigenstate (eigendistribution)*. If the particle is in an eigenstate of position, then its momentum is completely unknown. On the other hand, if the particle is in an eigenstate of momentum, then its position is completely unknown. In an eigenstate of momentum having a plane wave form, it can be shown that the wavelength is equal to h/p, where h is Planck's constant and p is the momentum of the eigenstate.

Step Potential

The potential in this case is given by:

$$V(x) = \begin{cases} 0, & x < 0, \\ V_0, & x \geq 0. \end{cases}$$

The solutions are superpositions of left- and right-moving waves:

$$\psi_1(x) = \frac{1}{\sqrt{k_1}}\left(A_{\rightarrow}e^{ik_1x} + A_{\leftarrow}e^{-ik_1x}\right) \quad x < 0$$

$$\psi_2(x) = \frac{1}{\sqrt{k_2}}\left(B_{\rightarrow}e^{ik_2x} + B_{\leftarrow}e^{-ik_2x}\right) \quad x > 0$$

where the wave vectors are related to the energy via

$$k_1 = \sqrt{2mE/\hbar^2}, \text{ and}$$

$$k_2 = \sqrt{2m(E - V_0)/\hbar^2}$$

with coefficients A and B determined from the boundary conditions and by imposing a continuous derivative on the solution.

Each term of the solution can be interpreted as an incident, reflected, or transmitted component of the wave, allowing the calculation of transmission and reflection coefficients. Notably, in contrast to classical mechanics, incident particles with energies greater than the potential step are partially reflected.

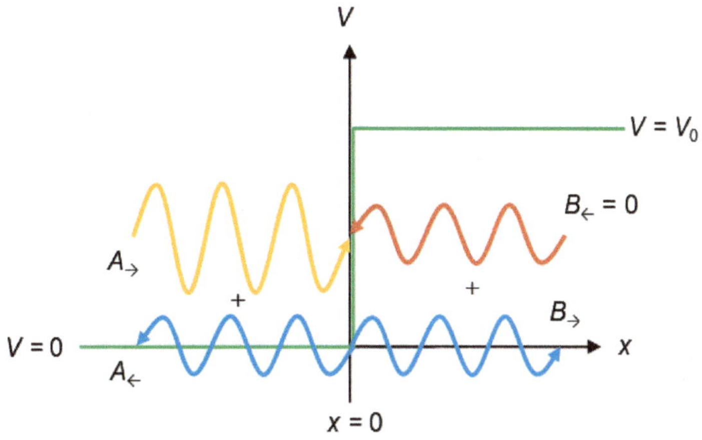

Scattering at a finite potential step of height V_0, shown in green. The amplitudes and direction of left- and right-moving waves are indicated. Yellow is the incident wave, blue are reflected and transmitted waves, red does not occur. $E > V_0$ for this figure.

Rectangular Potential Barrier

This is a model for the quantum tunneling effect which plays an important role in the performance of modern technologies such as flash memory and scanning tunneling microscopy. Quantum tunneling is central to physical phenomena involved in superlattices.

Particle in a Box

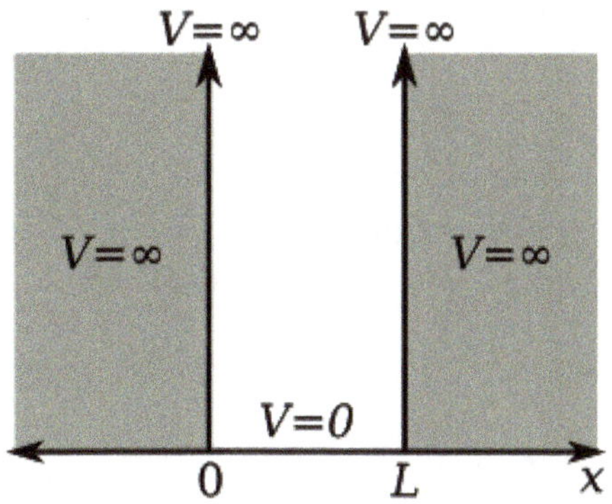

1-dimensional potential energy box (or infinite potential well)

The particle in a one-dimensional potential energy box is the most mathematically simple example where restraints lead to the quantization of energy levels. The box is defined as having zero potential energy everywhere *inside* a certain region, and infinite potential energy everywhere *outside* that region. For the one-dimensional case in the x direction, the time-independent Schrödinger equation may be written

$$-\frac{\hbar^2}{2m}\frac{d^2\psi}{dx^2} = E\psi.$$

With the differential operator defined by

$$\hat{p}_x = -i\hbar\frac{d}{dx}$$

the previous equation is evocative of the classic kinetic energy analogue,

$$\frac{1}{2m}\hat{p}_x^2 = E,$$

with state ψ in this case having energy E coincident with the kinetic energy of the particle.

The general solutions of the Schrödinger equation for the particle in a box are

$$\psi(x) = Ae^{ikx} + Be^{-ikx} \qquad E = \frac{\hbar^2 k^2}{2m}$$

or, from Euler's formula,

$$\psi(x) = C\sin kx + D\cos kx.$$

The infinite potential walls of the box determine the values of C, D, and k at x = 0 and x = L where ψ must be zero. Thus, at x = 0,

$$\psi(0) = 0 = C\sin 0 + D\cos 0 = D$$

and D = 0. At x = L,

$$\psi(L) = 0 = C\sin kL.$$

in which C cannot be zero as this would conflict with the Born interpretation. Therefore, since sin(kL) = 0, kL must be an integer multiple of π,

$$k = \frac{n\pi}{L} \qquad n = 1, 2, 3, \ldots.$$

The quantization of energy levels follows from this constraint on k, since

$$E = \frac{\hbar^2\pi^2 n^2}{2mL^2} = \frac{n^2 h^2}{8mL^2}.$$

Finite Potential Well

A finite potential well is the generalization of the infinite potential well problem to potential wells having finite depth.

The finite potential well problem is mathematically more complicated than the infinite particle-in-a-box problem as the wave function is not pinned to zero at the walls of the well. Instead, the wave function must satisfy more complicated mathematical boundary conditions as it is nonzero in regions outside the well.

Harmonic Oscillator

As in the classical case, the potential for the quantum harmonic oscillator is given by

$$V(x) = \frac{1}{2}m\omega^2 x^2.$$

This problem can either be treated by directly solving the Schrödinger equation, which is not trivial, or by using the more elegant "ladder method" first proposed by Paul Dirac. The eigenstates are given by

$$\psi_n(x) = \sqrt{\frac{1}{2^n\, n!}} \cdot \left(\frac{m\omega}{\pi\hbar}\right)^{1/4} \cdot e^{-\frac{m\omega x^2}{2\hbar}} \cdot H_n\left(\sqrt{\frac{m\omega}{\hbar}}x\right),$$

$$n = 0, 1, 2, \ldots.$$

where H_n are the Hermite polynomials

$$H_n(x) = (-1)^n e^{x^2} \frac{d^n}{dx^n}\left(e^{-x^2}\right),$$

and the corresponding energy levels are

$$E_n = \hbar\omega\left(n + \frac{1}{2}\right).$$

This is another example illustrating the quantification of energy for bound states.

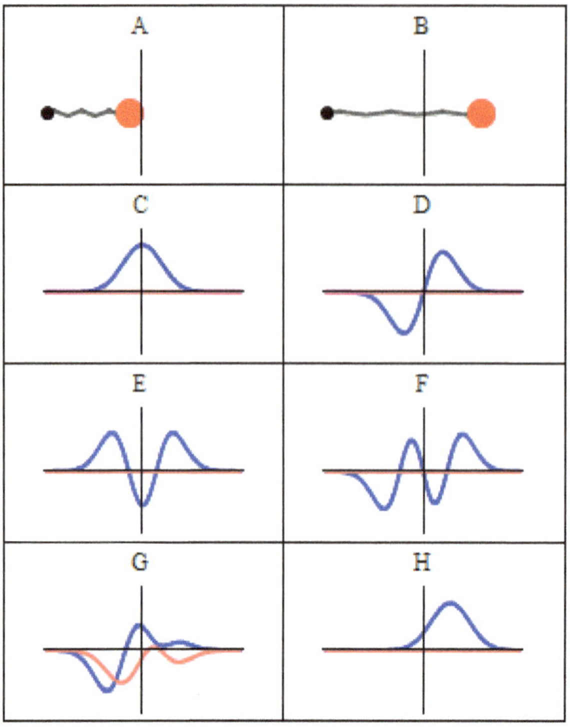

Some trajectories of a harmonic oscillator (i.e. a ball attached to a spring) in classical mechanics (A-B) and quantum mechanics (C-H). In quantum mechanics, the position of the ball is represented by a wave (called the wave function), with the real part shown in blue and the imaginary part shown in red. Some of the trajectories (such as C,D,E,and F) are standing waves (or "stationary states"). Each standing-wave frequency is proportional to a possible energy level of the oscillator. This "energy quantization" does not occur in classical physics, where the oscillator can have *any* energy.

REFERENCES.

1. "Quantum Mechanics". National Public Radio. Retrieved 22 June 2016.
2. Kuhn, Thomas S. The Structure of Scientific Revolutions. Fourth ed. Chicago; London: The University of Chicago Press, 2012. Print.
3. Feynman, Richard P. (1988). QED : the strange theory of light and matter (1st Princeton pbk., seventh printing with corrections. ed.). Princeton, N.J.: Princeton University Press. p. 10. ISBN 978-0691024172.
4. This result was published (in German) as Planck, Max (1901). "Ueber das Gesetz der Energieverteilung im Normalspectrum" (PDF). Ann. Phys. 309 (3): 553–63. Bibcode:1901AnP...309..553P. doi:10.1002/andp.19013090310.. English translation: "On the Law of Distribution of Energy in the Normal Spectrum".
5. Francis Weston Sears (1958). Mechanics, Wave Motion, and Heat. Addison-Wesley. p. 537.

6. "The Nobel Prize in Physics 1918". Nobel Foundation. Retrieved 2009-08-01.
7. Kragh, Helge (1 December 2000). "Max Planck: the reluctant revolutionary". PhysicsWorld.com.
8. Einstein, Albert (1905). "Über einen die Erzeugung und Verwandlung des Lichtes betreffenden heuristischen Gesichtspunkt (PDF). Annalen der Physik. 17 (6): 132–148. Bibcode:1905AnP...322..132E. doi:10.1002/andp.19053220607., translated into English as On a Heuristic Viewpoint Concerning the Production and Transformation of Light. The term photon was introduced in 1926.
9. Taylor, J. R.; Zafiratos, C. D.; Dubson, M. A. (2004). Modern Physics for Scientists and Engineers. Prentice Hall. pp. 127–9. ISBN 0-13-589789-0.
10. Stephen Hawking, The Universe in a Nutshell, Bantam, 2001.
11. Dicke and Wittke, Introduction to Quantum Mechanics, p. 12
12. NTRS.NASA.gov
13. Taylor, J. R.; Zafiratos, C. D.; Dubson, M. A. (2004). Modern Physics for Scientists and Engineers. Prentice Hall. pp. 147–8. ISBN 0-13-589789-0.
14. McEvoy, J. P.; Zarate, O. (2004). Introducing Quantum Theory. Totem \Books. pp. 70–89, especially p. 89. ISBN 1-84046-577-8.
15. World Book Encyclopedia, page 6, 2007.
16. Dicke and Wittke, Introduction to Quantum Mechanics, p. 10f.
17. J. P. McEvoy and Oscar Zarate (2004). Introducing Quantum Theory. Totem Books. p. 110f. ISBN 1-84046-577-8.
18. Aczel, Amir D., Entanglement, p. 51f. (Penguin, 2003) ISBN 978-1-5519-2647-6
19. J. P. McEvoy and Oscar Zarate (2004). Introducing Quantum Theory. Totem Books. p. 114. ISBN 1-84046-577-8.
20. Zettili, Nouredine (2009). Quantum Mechanics: Concepts and Applications. John Wiley and Sons. pp. 26–27. ISBN 0470026782.
21. Selleri, Franco (2012). Wave-Particle Duality. Springer Science and Business Media. p. 41. ISBN 1461533325.
22. Podgorsak, Ervin B. (2013). Compendium to Radiation Physics for Medical Physicists. Springer Science and Business Media. p. 88. ISBN 3642201865.
23. Halliday, David; Resnick, Robert (2013). Fundamentals of Physics, 10th Ed. John Wiley and Sons. p. 1272. ISBN 1118230612.
24. Myers, Rusty L. (2006). The Basics of Physics. Greenwood Publishing Group. p. 172. ISBN 0313328579.
25. Introducing Quantum Theory, p. 87
26. Van der Waerden, B. L. (1967). Sources of Quantum Mechanics. Mineola, New York: Dover Publications. pp. 261–276. Received July 29, 1925 See Werner Heisenberg's paper, Quantum-Theoretical Re-interpretation of Kinematic and Mechanical Relations pp. 261-276
27. Nobel Prize Organization. Erwin Schrödinger - Biographical. Retrieved 28 March

2014. His great discovery, Schrödinger's wave equation, was made at the end of this epoch-during the first half of 1926.

28. Schrodinger Equation (Physics), Encyclopædia Britannica
29. Erwin Schrödinger, The Present Situation in Quantum Mechanics, p. 9. This translation was originally published in Proceedings of the American Philosophical Society, 124, 323-38, and then appeared as Section I.11 of Part I of Quantum Theory and Measurement (J.A. Wheeler and W.H. Zurek, eds., Princeton university Press, New Jersey 1983). This paper can be downloaded from http://www.tu-harburg.de/rzt/rzt/it/QM/cat.html.
30. Heisenberg, W. (1955). The development of the interpretation of the quantum theory, pp. 12–29 in Niels Bohr and the Development of Physics: Essays dedicated to Niels Bohr on the occasion of his seventieth birthday, edited by Pauli, W. with the assistance of Rosenfeld, L. and Weisskopf, V., Pergamon, London, p. 13: the single quantum jump ... is factual in nature.
31. W. Moore, Schrödinger: Life and Thought, Cambridge University Press (1989), p. 222. See p. 227 for Schrödinger's own words.
32. Heisenberg's Nobel Prize citation
33. Heisenberg first published his work on the uncertainty principle in the leading German physics journal Zeitschrift für Physik: Heisenberg, W. (1927). Über den anschaulichen Inhalt der quantentheoretischen Kinematik und Mechanik. Z. Phys. 43 (3–4): 172–198. Bibcode:1927ZPhy...43..172H. doi:10.1007/BF01397280.
34. Nobel Prize in Physics presentation speech, 1932
35. Uncertainty principle, Encyclopædia Britannica
36. Pauling, Linus (1960). The Nature of the Chemical Bond (3rd ed.). Itahca, New York: Cornell University Press. p. 47. ISBN 0-8014-0333-2. Retrieved 1 March 2016.
37. Orbital (chemistry and physics), Encyclopædia Britannica
38. E. Schrödinger, Proceedings of the Cambridge Philosophical Society, 31 (1935), p. 555, says: When two systems, of which we know the states by their respective representation, enter into a temporary physical interaction due to known forces between them and when after a time of mutual influence the systems separate again, then they can no longer be described as before, viz., by endowing each of them with a representative of its own. I would not call that one but rather the characteristic trait of quantum mechanics.
39. Quantum Nonlocality and the Possibility of Superluminal Effects, John G. Cramer, npl.washington.edu
40. Mechanics, Merriam-Webster Online Dictionary
41. Field, Encyclopædia Britannica
42. Richard Hammond, The Unknown Universe, New Page Books, 2008. ISBN 978-1-60163-003-2

43. The Physical World website
44. The Nobel Prize in Physics 1933. Nobel Foundation. Retrieved 2007-11-24.
45. Durrani, Z. A. K.; Ahmed, H. (2008). Vijay Kumar, ed. Nanosilicon. Elsevier. p. 345. ISBN 978-0-08-044528-1
46. Matson, John. "What Is Quantum Mechanics Good for?". Scientific American. Retrieved 18 May 2016.
47. Max Born & Emil Wolf, Principles of Optics, 1999, Cambridge University Press
48. Mehra, J.; Rechenberg, H. (1982). The historical development of quantum theory. New York: Springer-Verlag. ISBN 0387906428.
49. Kragh, Helge (2002). Quantum Generations: A History of Physics in the Twentieth Century. Princeton University Press. p. 58. ISBN 0-691-09552-3. Extract of page 58
50. Ben-Menahem, Ari (2009). Historical Encyclopedia of Natural and Mathematical Sciences, Volume 1. Springer. p. 3678. ISBN 3540688315. Extract of page 3678
51. E Arunan (2010). "Peter Debye" (PDF). Resonance (journal). Indian Academy of Sciences. 15 (12).
52. Kuhn, T. S. (1978). Black-body theory and the quantum discontinuity 1894-1912. Oxford: Clarendon Press. ISBN 0195023838.
53. Kragh, Helge (1 December 2000), Max Planck: the reluctant revolutionary, PhysicsWorld.com
54. Einstein, A. (1905). "Über einen die Erzeugung und Verwandlung des Lichtes betreffenden heuristischen Gesichtspunkt [On a heuristic point of view concerning the production and transformation of light]. Annalen der Physik. 17 (6): 132–148. Bibcode:1905AnP...322..132E. doi:10.1002/andp.19053220607. Reprinted in The collected papers of Albert Einstein, John Stachel, editor, Princeton University Press, 1989, Vol. 2, pp. 149-166, in German; see also Einstein's early work on the quantum hypothesis, ibid. pp. 134-148.
55. van Hove, Leon (1958). Von Neumann's contributions to quantum mechanics (PDF). Bulletin of the American Mathematical Society. 64 (3): Part2:95–99. doi:10.1090/s0002-9904-1958-10206-2.
56. Feynman, Richard. The Feynman Lectures on Physics III 21-4. California Institute of Technology. Retrieved 2015-11-24. ...it was long believed that the wave function of the Schrödinger equation would never have a macroscopic representation analogous to the macroscopic representation of the amplitude for photons. On the other hand, it is now realized that the phenomena of superconductivity presents us with just this situation.
57. Richard Packard (2006) Berkeley Experiments on Superfluid Macroscopic Quantum Effects Archived November 25, 2015, at the Wayback Machine. accessdate=2015-11-24
58. Quantum - Definition and More from the Free Merriam-Webster Dictionary. Merriam-webster.com. Retrieved 2012-08-18.

59. Thall, Edwin. Thall's History of Quantum Mechanics. Florida Community College at Jacksonville. Archived from the original on October 7, 2009. Retrieved May 23, 2009.
60. ysfine.com. ysfine.com. Retrieved 11 September 2015.
61. QUANTUM MECHANICS. geocities.com. 2009-10-26. Archived from the original on 2009-10-26. Retrieved 2016-06-13.
62. P.A.M. Dirac, The Principles of Quantum Mechanics, Clarendon Press, Oxford, 1930.
63. D. Hilbert Lectures on Quantum Theory, 1915–1927
64. J. von Neumann, Mathematische Grundlagen der Quantenmechanik, Springer, Berlin, 1932 (English translation: Mathematical Foundations of Quantum Mechanics, Princeton University Press, 1955).
65. H.Weyl The Theory of Groups and Quantum Mechanics, 1931 (original title: Gruppentheorie und Quantenmechanik).
66. Dirac, P.A.M. (1958). The Principles of Quantum Mechanics, 4th edition, Oxford University Press, Oxford UK, p. ix: For this reason I have chosen the symbolic method, introducing the representatives later merely as an aid to practical calculation.
67. Greiner, Walter; Müller, Berndt (1994). Quantum Mechanics Symmetries, Second edition. Springer-Verlag. p. 52. ISBN 3-540-58080-8., Chapter 1, p. 52
68. Heisenberg - Quantum Mechanics, 1925–1927: The Uncertainty Relations. Aip.org. Retrieved 2012-08-18.
69. Greenstein, George; Zajonc, Arthur (2006). The Quantum Challenge: Modern Research on the Foundations of Quantum Mechanics, Second edition. Jones and Bartlett Publishers, Inc. p. 215. ISBN 0-7637-2470-X., Chapter 8, p. 215
70. [Abstract] Visualization of Uncertain Particle Movement. Actapress.com. Retrieved 2012-08-18.
71. Hirshleifer, Jack (2001). The Dark Side of the Force: Economic Foundations of Conflict Theory. Campbridge University Press. p. 265. ISBN 0-521-80412-4., Chapter , p.
72. dict.cc dictionary :: eigen :: German-English translation. dict.cc. Retrieved 11 September 2015.
73. Topics: Wave-Function Collapse. Phy.olemiss.edu. 2012-07-27. Retrieved 2012-08-18.
74. Collapse of the wave-function. Farside.ph.utexas.edu. Retrieved 2012-08-18.
75. Determinism and Naive Realism : philosophy. Reddit.com. 2009-06-01. Retrieved 2012-08-18.
76. Michael Trott. Time-Evolution of a Wavepacket in a Square Well — Wolfram Demonstrations Project. Demonstrations.wolfram.com. Retrieved 2010-10-15.
77. Michael Trott. Time Evolution of a Wavepacket In a Square Well. Demonstrations.wolfram.com. Retrieved 2010-10-15.

78. Mathews, Piravonu Mathews; Venkatesan, K. (1976). A Textbook of Quantum Mechanics. Tata McGraw-Hill. p. 36. ISBN 0-07-096510-2., Chapter 2, p. 36
79. Wave Functions and the Schrödinger Equation (PDF). Retrieved 2010-10-15.
80. Rechenberg, Helmut (1987). Erwin Schrödinger and the creation of wave mechanics (PDF). Acta Physica Polonica B. 19 (8): 683–695. Retrieved 13 June 2016.
81. Nancy Thorndike Greenspan, The End of the Certain World: The Life and Science of Max Born (Basic Books, 2005), pp. 124-8 and 285-6.
82. http://ocw.usu.edu/physics/classical-mechanics/pdf_lectures/06.pdf
83. The Nobel Prize in Physics 1979. Nobel Foundation. Retrieved 2010-02-16.
84. Carl M. Bender; Daniel W. Hook; Karta Kooner (2009-12-31). Complex Elliptic Pendulum. arXiv:1001.0131 [hep-th].
85. See, for example, Precision tests of QED. The relativistic refinement of quantum mechanics known as quantum electrodynamics (QED) has been shown to agree with experiment to within 1 part in 108 for some atomic properties.
86. Tipler, Paul; Llewellyn, Ralph (2008). Modern Physics (5 ed.). W. H. Freeman and Company. pp. 160–161. ISBN 978-0-7167-7550-8.
87. Quantum mechanics course iwhatisquantummechanics. Scribd.com. 2008-09-14. Retrieved 2012-08-18.
88. A. Einstein, B. Podolsky, and N. Rosen, Can quantum-mechanical description of physical reality be considered complete? Phys. Rev. 47 777 (1935).
89. N. P. Landsman (June 13, 2005). Between classical and quantum(PDF). Retrieved 2012-08-19. Handbook of the Philosophy of Science Vol. 2: Philosophy of Physics (eds. John Earman & Jeremy Butterfield).
90. (see macroscopic quantum phenomena, Bose–Einstein condensate, and Quantum machine)
91. Atomic Properties. Academic.brooklyn.cuny.edu. Retrieved 2012-08-18.
92. http://assets.cambridge.org/97805218/29526/excerpt/9780521829526_excerpt.pdf
93. Born, M., Heisenberg, W., Jordan, P. (1926). Z. Phys. 35: 557–615. Translated as ‹On quantum mechanics II', pp. 321–385 in Van der Waerden, B.L. (1967), Sources of Quantum Mechanics, North-Holland, Amsterdam, The basic difference between the theory proposed here and that used hitherto ... lies in the characteristic kinematics ..., p. 385.
94. Dirac, P.A.M. (1930/1958). The Principles of Quantum Mechanics, fourth edition, Oxford University Press, Oxford UK, p. 5: A question about what will happen to a particular photon under certain conditions is not really very precise. To make it precise one must imagine some experiment performed having a bearing on the question, and enquire what will be the result of the experiment. Only questions about the results of experiments have a real significance and it is only such questions that theoretical physics has to consider.
95. Bohr, N. (1939). The Causality Problem in Atomic Physics, in New Theories

in Physics, Conference organized in collaboration with the International Union of Physics and the Polish Intellectual Co-operation Committee, Warsaw, May 30th – June 3rd 1938, International Institute of Intellectual Co-operation, Paris, 1939, pp. 11–30, reprinted in Niels Bohr, Collected Works, volume 7 (1933 – 1958) edited by J. Kalckar, Elsevier, Amsterdam, ISBN 0-444-89892-1, pp. 303–322. The essential lesson of the analysis of measurements in quantum theory is thus the emphasis on the necessity, in the account of the phenomena, of taking the whole experimental arrangement into consideration, in complete conformity with the fact that all unambiguous interpretation of the quantum mechanical formalism involves the fixation of the external conditions, defining the initial state of the atomic system and the character of the possible predictions as regards subsequent observable properties of that system. Any measurement in quantum theory can in fact only refer either to a fixation of the initial state or to the test of such predictions, and it is first the combination of both kinds which constitutes a well-defined phenomenon.

96. Bohr, N. (1948). On the notions of complementarity and causality, Dialectica 2: 312–319. As a more appropriate way of expression, one may advocate limitation of the use of the word phenomenon to refer to observations obtained under specified circumstances, including an account of the whole experiment.
97. Ludwig, G. (1987). An Axiomatic Basis for Quantum Mechanics, volume 2, Quantum Mechanics and Macrosystems, translated by K. Just, Springer, Berlin, ISBN 978-3-642-71899-1, Chapter XIII, Special Structures in Preparation and Registration Devices, §1, Measurement chains, p. 132.
98. Heisenberg, W. (1927). Über den anschaulichen Inhalt der quantentheoretischen Kinematik und Mechanik, Z. Phys. 43: 172–198. Translation as ‹The actual content of quantum theoretical kinematics and mechanics’ here , But in the rigorous formulation of the law of causality, — If we know the present precisely, we can calculate the future — it is not the conclusion that is faulty, but the premise.
99. Green, H.S. (1965). Matrix Mechanics, with a foreword by Max Born, P. Noordhoff Ltd, Groningen. It is not possible, therefore, to provide ‹initial conditions’ for the prediction of the behaviour of atomic systems, in the way contemplated by classical physics. This is accepted by quantum theory, not merely as an experimental difficulty, but as a fundamental law of nature, p. 32.
100. Rosenfeld, L. (1957). Misunderstandings about the foundations of quantum theory, pp. 41–45 in Observation and Interpretation, edited by S. Körner, Butterworths, London. A phenomenon is therefore a process (endowed with the characteristic quantal wholeness) involving a definite type of interaction between the system and the apparatus.
101. Dirac, P.A.M. (1973). Development of the physicist’s conception of nature, pp. 1–55 in The Physicist’s Conception of Nature, edited by J. Mehra, D. Reidel, Dordrecht, ISBN 90-277-0345-0, p. 5: That led Heisenberg to his really masterful step forward, resulting in the new quantum mechanics. His idea was to build up a theory entirely in terms of quantities referring to two states.
102. Born, M. (1927). Physical aspects of quantum mechanics, Nature 119: 354–357,

These probabilities are thus dynamically determined. But what the system actually does is not determined ...

103. Messiah, A. (1961). Quantum Mechanics, volume 1, translated by G.M. Temmer from the French Mécanique Quantique, North-Holland, Amsterdam, p. 157.
104. Bohr, N. (1928). The Quantum postulate and the recent development of atomic theory, Nature 121: 580–590.
105. Heisenberg, W. (1930). The Physical Principles of the Quantum Theory, translated by C. Eckart and F.C. Hoyt, University of Chicago Press.
106. Goldstein, H. (1950). Classical Mechanics, Addison-Wesley, ISBN 0-201-02510-8.
107. There is as yet no logically consistent and complete relativistic quantum field theory., p. 4. — V. B. Berestetskii, E. M. Lifshitz, L P Pitaevskii (1971). J. B. Sykes, J. S. Bell (translators). Relativistic Quantum Theory 4, part I. Course of Theoretical Physics (Landau and Lifshitz) ISBN 0-08-016025-5
108. Stephen Hawking; Gödel and the end of physics. cam.ac.uk. Retrieved 11 September 2015.
109. The Nature of Space and Time. google.com. Retrieved 11 September 2015.
110. Tatsumi Aoyama; Masashi Hayakawa; Toichiro Kinoshita; Makiko Nio (2012). Tenth-Order QED Contribution to the Electron g-2 and an Improved Value of the Fine Structure Constant. Physical Review Letters. 109 (11): 111807. arXiv:1205.5368v2 . Bibcode:2012PhRvL.109k1807A. doi:10.1103/PhysRevLett.109.111807.
111. Parker, B. (1993). Overcoming some of the problems. pp. 259–279.
112. The Character of Physical Law (1965) Ch. 6; also quoted in The New Quantum Universe (2003), by Tony Hey and Patrick Walters
113. Weinberg, S. Collapse of the State Vector, Phys. Rev. A 85, 062116 (2012).
114. Harrison, Edward (16 March 2000). Cosmology: The Science of the Universe. Cambridge University Press. p. 239. ISBN 978-0-521-66148-5.
115. Action at a Distance in Quantum Mechanics (Stanford Encyclopedia of Philosophy). Plato.stanford.edu. 2007-01-26. Retrieved 2012-08-18.
116. Everett's Relative-State Formulation of Quantum Mechanics (Stanford Encyclopedia of Philosophy). Plato.stanford.edu. Retrieved 2012-08-18.
117. The Transactional Interpretation of Quantum Mechanics by John Cramer. Reviews of Modern Physics 58, 647-688, July (1986)
118. See, for example, the Feynman Lectures on Physics for some of the technological applications which use quantum mechanics, e.g., transistors (vol III, pp. 14–11 ff), integrated circuits, which are follow-on technology in solid-state physics (vol II, pp. 8–6), and lasers (vol III, pp. 9–13).
119. Pauling, Linus; Wilson, Edgar Bright (1985-03-01). Introduction to Quantum Mechanics with Applications to Chemistry. ISBN 9780486648712. Retrieved 2012-08-18.

120. Chen, Xie; Gu, Zheng-Cheng; Wen, Xiao-Gang (2010). Local unitary transformation, long-range quantum entanglement, wave function renormalization, and topological order. Phys. Rev. B. 82: 155138. arXiv:1004.3835 . Bibcode:2010PhRvB..82o5138C. doi:10.1103/physrevb.82.155138.

121. Anderson, Mark (2009-01-13). Is Quantum Mechanics Controlling Your Thoughts? | Subatomic Particles. DISCOVER Magazine. Retrieved 2012-08-18.

122. Quantum mechanics boosts photosynthesis. physicsworld.com. Retrieved 2010-10-23.

123. Davies, P. C. W.; Betts, David S. (1984). Quantum Mechanics, Second edition. Chapman and Hall. p. 79. ISBN 0-7487-4446-0., Chapter 6, p. 79

124. Baofu, Peter (2007-12-31). The Future of Complexity: Conceiving a Better Way to Understand Order and Chaos. ISBN 9789812708991. Retrieved 2012-08-18.

125. Derivation of particle in a box, chemistry.tidalswan.com

CHAPTER 2

Quantum Theory from The Years 1900–1925

Chapter Contents

- **Introduction**
- **Basic Principles**
- **Examples**
- **De Broglie Waves**
- **Kramers Transition Matrix**

Bohr's 1913 model of the hydrogen atom really starts the story. Nineteenth and early twentieth century experimentalists heaped up data on the spectra of atoms. Besides the plain old spectra of hydrogen, helium, sodium, etc.,

INTRODUCTION

The old quantum theory is a collection of results from the years 1900–1925 which predate modern quantum mechanics. The theory was never complete or self-consistent, but was a set of heuristic prescriptions which are now understood to be the first quantum corrections to classical mechanics. The Bohr model was the focus of study, and Arnold Sommerfeld made a crucial contribution by quantizing the z-component of the angular momentum, which in the old quantum era was called *space quantization* (Richtungsquantelung). This allowed the orbits of the electron to be ellipses instead of circles, and introduced the concept of quantum degeneracy. The theory would have correctly explained the Zeeman effect, except for the issue of electron spin.

The main tool was Bohr–Sommerfeld quantization, a procedure for selecting out certain discrete set of states of a classical integrable motion as allowed states. These are like the allowed orbits of the Bohr model of the atom; the system can only be in one of these states and not in any states in between.

BASIC PRINCIPLES

The basic idea of the old quantum theory is that the motion in an atomic system is quantized, or discrete. The system obeys classical mechanics except that not every motion is allowed, only those motions which obey the *old quantum condition*:

$$\oint_{H(p,q)=E} p_i \, dq_i = n_i h$$

where the p_i are the momenta of the system and the q_i are the corresponding coordinates. The quantum numbers n_i and the integral is taken over one period of the motion at constant energy (as described by the Hamiltonian). The integral is an area in phase space, which is a quantity called the action and is quantized in units of Planck's constant. For this reason, Planck's constant was often called the *quantum of action*.

In order for the old quantum condition to make sense, the classical motion must be separable, meaning that there are separate coordinates q_i in terms of which the motion is periodic. The periods of the different motions do not have to be the same, they can even be incommensurate, but there must be a set of coordinates where the motion decomposes in a multi-periodic way.

The motivation for the old quantum condition was the correspondence principle, complemented by the physical observation that the quantities which are quantized must be adiabatic invariants. Given Planck's quantization rule for the harmonic oscillator, either condition determines the correct classical quantity to quantize in a general system up to an additive constant.

EXAMPLES

Thermal Properties of the Harmonic Oscillator

The simplest system in the old quantum theory is the harmonic oscillator, whose Hamiltonian is:

$$H = \frac{p^2}{2m} + \frac{m\omega^2 q^2}{2}.$$

The old quantum theory yields a recipe for the quantization of the energy levels of the harmonic oscillator, which, when combined with the Boltzmann probability distribution of thermodynamics, yields the correct expression for the stored energy and specific heat of a quantum oscillator both at low and at ordinary temperatures. Applied as a model for the specific heat of solids, this resolved a discrepancy in pre-quantum thermodynamics that had troubled 19th-century scientists. Let us now describe this.

The level sets of H are the orbits, and the quantum condition is that the area enclosed by an orbit in phase space is an integer. It follows that the energy is quantized according to the Planck rule:

$$E = n\hbar\omega,$$

A result which was known well before, and used to formulate the old quantum condition. This result differs by $\frac{1}{2}\hbar\omega$ from the results found with the help of quantum mechanics. This constant is neglected in the derivation of the *old quantum theory*, and its value can not be determined using it.

The thermal properties of a quantized oscillator may be found by averaging the energy in each of the discrete states assuming that they are occupied with a Boltzmann weight:

$$U = \frac{\sum_n \hbar\omega n e^{-\beta n\hbar\omega}}{\sum_n e^{-\beta n\hbar\omega}} = \frac{\hbar\omega e^{-\beta\hbar\omega}}{1 - e^{-\beta\hbar\omega}}, \quad \text{where } \beta = \frac{1}{kT},$$

kT is Boltzmann constant times the absolute temperature, which is the temperature as measured in more natural units of energy. The quantity β is more fundamental in thermodynamics than the temperature, because it is the thermodynamic potential associated to the energy.

From this expression, it is easy to see that for large values of β for very low temperatures, the average energy U in the Harmonic oscillator approaches zero very quickly, exponentially fast. The reason is that kT is the typical energy of random motion at temperature T, and when this is smaller than $\hbar\omega$, there is not enough energy to give the oscillator even one quantum of energy. So the oscillator stays in its ground state, storing next to no energy at all.

This means that at very cold temperatures, the change in energy with respect to beta, or equivalently the change in energy with respect to temperature, is also exponentially small. The change in energy with respect to temperature is the specific heat, so the specific heat is exponentially small at low temperatures, going to zero like

$$\exp(-\hbar\omega/kT)$$

At small values of β, at high temperatures, the average energy U is equal to 1/β = kT This reproduces the equipartition theorem of classical thermodynamics: every harmonic oscillator at temperature T has energy kT on average. This means that the specific heat of an oscillator is constant in classical mechanics and equal to k. For a collection of atoms connected by springs, a reasonable model of a solid, the total specific heat is equal to the total number of oscillators times k. There are overall three oscillators for each atom, corresponding to the three possible directions of independent oscillations in three dimensions. So the specific heat of a classical solid is always 3k per atom, or in chemistry units, 3R per mole of atoms.

Monatomic solids at room temperatures have approximately the same specific heat of 3k per atom, but at low temperatures they don't. The specific heat is smaller at colder temperatures, and it goes to zero at absolute zero. This is true for all material systems, and this observation is called the third law of thermodynamics. Classical mechanics cannot explain the third law, because in classical mechanics the specific heat is independent of the temperature.

This contradiction between classical mechanics and the specific heat of cold materials was noted by James Clerk Maxwell in the 19th century, and remained a deep puzzle for those who advocated an atomic theory of matter. Einstein resolved this problem in 1906 by proposing that atomic motion is quantized. This was the first application of quantum theory to mechanical systems. A short while later, Peter Debye gave a quantitative theory of solid specific heats in terms of quantized oscillators with various frequencies (see Einstein solid and Debye model).

One-Dimensional Potential: U=0

One-dimensional problems are easy to solve. At any energy E, the value of the momentum p is found from the conservation equation:

$$\sqrt{2m(E - V(q))} = p$$

which is integrated over all values of q between the classical *turning points*, the places where the momentum vanishes. The integral is easiest for a *particle in a box* of length L, where the quantum condition is:

$$2\int_0^L p\,dq = nh$$

which gives the allowed momenta:

$$p = \frac{nh}{2L}$$

and the energy levels

$$E_n = \frac{p^2}{2m} = \frac{n^2h^2}{8mL^2}$$

One-Dimensional Potential: U=Fx

Another easy case to solve with the old quantum theory is a linear potential on the positive halfline, the constant confining force F binding a particle to an impenetrable wall. This case is much more difficult in the full quantum mechanical treatment, and unlike the other examples, the semiclassical answer here is not exact but approximate, becoming more accurate at large quantum numbers.

$$2\int_0^{\frac{E}{F}} \sqrt{2m(E-Fx)}\,dx = nh$$

so that the quantum condition is

$$\frac{4}{3}\sqrt{2m}\frac{E^{3/2}}{F} = nh$$

which determines the energy levels,

$$E_n = \left(\frac{3nhF}{4\sqrt{2m}}\right)^{2/3}$$

In the specific case F=mg, the particle is confined by the gravitational potential of the earth and the "wall" here is the surface of the earth.

One-Dimensional Potential: U=(1/2)kx^2

This case is also easy to solve, and the semiclassical answer here agrees with the quantum one to within the ground-state energy. Its quantization-condition integral is

$$2\int_{-\sqrt{\frac{2E}{k}}}^{\sqrt{\frac{2E}{k}}} \sqrt{2m\left(E-\frac{1}{2}kx^2\right)}\,dx = nh$$

with solution

$$E = n\frac{h}{2\pi}\sqrt{\frac{k}{m}} = n\hbar\omega$$

for oscillation angular frequency w, as before.

Rotator

Another simple system is the rotator. A rotator consists of a mass *M* at the end of a massless rigid rod of length *R* and in two dimensions has the Lagrangian:

$$L = \frac{MR^2}{2}\dot{\theta}^2$$

which determines that the angular momentum J conjugate to θ, the polar angle, $J=MR^2\theta$. The old quantum condition requires that J multiplied by the period of θ is an integer multiple of Planck's constant:

$$2\pi J = nh$$

the angular momentum to be an integer multiple of $\hbar$. In the Bohr model, this restriction imposed on circular orbits was enough to determine the energy levels.

In three dimensions, a rigid rotator can be described by two angles – θ and ϕ where θ is the inclination relative to an arbitrarily chosen *z*-axis while ϕ is the rotator angle in the projection to the *x–y* plane. The kinetic energy is again the only contribution to the Lagrangian:

$$L = \frac{MR^2}{2}\dot{\theta}^2 + \frac{MR^2}{2}(\sin(\theta)\dot{\phi})^2$$

And the conjugate momenta are $p_\theta = \dot{\theta}$ and $p_\phi = \sin(\theta)^2\dot{\phi}$. The equation of motion for is trivial: is a constant:

$$p_\phi = l_\phi$$

which is the z-component of the angular momentum. The quantum condition demands that the integral of the constant l_ϕ as ϕ varies from 0 to 2π is an integer multiple of h:

$$l_\phi = m\hbar$$

And *m* is called the magnetic quantum number, because the *z* component of the angular momentum is the magnetic moment of the rotator along the *z* direction in the case where the particle at the end of the rotator is charged.

Since the three-dimensional rotator is rotating about an axis, the total angular momentum should be restricted in the same way as the two-dimensional rotator. The two quantum conditions restrict the total angular momentum and the *z*-component of the angular momentum to be the integers *l*,*m*. This condition is reproduced in modern quantum mechanics, but in the era of the old quantum theory it led to a paradox: how can the orientation of the angular momentum relative to the arbitrarily chosen *z*-axis be quantized? This seems to pick out a direction in space.

This phenomenon, the quantization of angular momentum about an axis, was given the name *space quantization*, because it seemed incompatible with rotational invariance. In modern quantum mechanics, the angular momentum is quantized the same way, but the discrete states of definite angular momentum in any one orientation are quantum superpositions of the states in other orientations, so that the process of quantization does not pick out a preferred axis. For this reason, the name "space quantization" fell out of favor, and the same phenomenon is now called the quantization of angular momentum.

Hydrogen Atom

The angular part of the hydrogen atom is just the rotator, and gives the quantum numbers *l* and *m*. The only remaining variable is the radial coordinate, which executes a periodic one-dimensional potential motion, which can be solved.

For a fixed value of the total angular momentum *L*, the Hamiltonian for a classical Kepler problem is (the unit of mass and unit of energy redefined to absorb two constants):

$$H = \frac{p^2}{2} + \frac{l^2}{2r^2} - \frac{1}{r}.$$

Fixing the energy to be (a negative) constant and solving for the radial momentum p, the quantum condition integral is:

$$2\oint\sqrt{2E-\frac{l^2}{r^2}+\frac{2}{r}}\,dr=kh$$

which can be solved with the method of residues, and gives a new quantum number k which determines the energy in combination with l. The energy is:

$$E=-\frac{1}{2(k+l)^2}$$

and it only depends on the sum of k and l, which is the *principal quantum number n*. Since k is positive, the allowed values of l for any given n are no bigger than n. The energies reproduce those in the Bohr model, except with the correct quantum mechanical multiplicities, with some ambiguity at the extreme values.

The semiclassical hydrogen atom is called the Sommerfeld model, and its orbits are ellipses of various sizes at discrete inclinations. The Sommerfeld model predicted that the magnetic moment of an atom measured along an axis will only take on discrete values, a result which seems to contradict rotational invariance but which was confirmed by the Stern–Gerlach experiment. This Bohr–Sommerfeld theory is a significant step in the development of quantum mechanics. It also describes the possibility of atomic energy levels being split by a magnetic field (called the Zeeman effect).

Relativistic Orbit

Arnold Sommerfeld derived the relativistic solution of atomic energy levels. We will start this derivation with the relativistic equation for energy in the electric potential

$$W=m_0c^2\left(\frac{1}{\sqrt{1-\frac{v^2}{c^2}}}-1\right)-k\frac{Ze^2}{r}$$

After substitution $u=\frac{1}{r}$ we get

$$\frac{1}{\sqrt{1-\frac{v^2}{c^2}}}=1+\frac{W}{m_0c^2}+k\frac{Ze^2}{m_0c^2}u$$

For momentum $p_r=m\dot{r}$, $p_\varphi=mr^2\dot{\varphi}$ and their ratio $\frac{p_r}{p_\varphi}=-\frac{du}{d\varphi}$ the equation of motion is (see Binet equation) we will obtain energies

$$\frac{d^2u}{d\varphi^2}=-\left(1-k^2\frac{Z^2e^4}{c^2p_\varphi^2}\right)u+\frac{m_0kZe^2}{p_\varphi^2}\left(1+\frac{W}{m_0c^2}\right)=-\omega_0^2u+K$$

with solution

$$u = \frac{1}{r} = K + A\cos\omega_0\varphi$$

The angular shift of periapsis per revolution is given by

$$\varphi_s = 2\pi\left(\frac{1}{\omega_0} - 1\right) \approx 4\pi^3 k^2 \frac{Z^2 e^4}{c^2 n_\varphi^2 h^2}$$

With the quantum conditions

$$\oint p_\varphi \, d\varphi = 2\pi p_\varphi = n_\varphi h$$

and

$$\oint p_r \, dr = p_\varphi \oint \left(\frac{1}{r}\frac{dr}{d\varphi}\right)^2 d\varphi = n_r h$$

we will obtain energies

$$\frac{W}{m_0 c^2} = \left(1 + \frac{\alpha^2 Z^2}{(n_r + \sqrt{n_\varphi^2 - \alpha^2 Z^2})^2}\right)^{-1/2} - 1$$

Where α is the fine-structure constant. This solution (using substitutions for quantum numbers) is equivalent to the solution of the Dirac equation.Nevertheless, both solutions fail to predict the Lamb shifts.

DE BROGLIE WAVES

In 1905, Einstein noted that the entropy of the quantized electromagnetic field oscillators in a box is, for short wavelength, equal to the entropy of a gas of point particles in the same box. The number of point particles is equal to the number of quanta. Einstein concluded that the quanta could be treated as if they were localizable objects (see page 139/140), particles of light, and named them photons.

Einstein's theoretical argument was based on thermodynamics, on counting the number of states, and so was not completely convincing. Nevertheless, he concluded that light had attributes of both waves and particles, more precisely that an electromagnetic standing wave with frequency ω with the quantized energy:

$$E = n\hbar\omega$$

should be thought of as consisting of n photons each with an energy Einstein could not describe how the photons were related to the wave.

The photons have momentum as well as energy, and the momentum had to be $\hbar\omega$. k is the wavenumber of the electromagnetic wave. This is required by relativity, because the momentum and energy form a four-vector, as do the frequency and wave-number.

In 1924, as a PhD candidate, Louis de Broglie proposed a new interpretation of the quantum condition. He suggested that all matter, electrons as well as photons, are described by waves obeying the relations.

$$p = \hbar k$$

or, expressed in terms of wavelength λ instead,

$$p = \frac{h}{\lambda}$$

He then noted that the quantum condition:

$$\int p\,dx = \hbar \int k\,dx = 2\pi\hbar n$$

counts the change in phase for the wave as it travels along the classical orbit, and requires that it be an integer multiple of 2π Expressed in wavelengths, the number of wavelengths along a classical orbit must be an integer. This is the condition for constructive interference, and it explained the reason for quantized orbits—the matter waves make standing waves only at discrete frequencies, at discrete energies.

For example, for a particle confined in a box, a standing wave must fit an integer number of wavelengths between twice the distance between the walls. The condition becomes:

$$n\lambda = 2L$$

so that the quantized momenta are:

$$p = \frac{nh}{2L}$$

reproducing the old quantum energy levels.

This development was given a more mathematical form by Einstein, who noted that the phase function for the waves: $\theta(J, x)$ in a mechanical system should be identified with the solution to the Hamilton–Jacobi equation, an equation which even Hamilton considered to be the short-wavelength limit of wave mechanics.

These ideas led to the development of the Schrödinger equation.

KRAMERS TRANSITION MATRIX

The old quantum theory was formulated only for special mechanical systems which could be separated into action angle variables which were periodic. It did not deal with the emission and absorption of radiation. Nevertheless, Hendrik Kramers was able to find heuristics for describing how emission and absorption should be calculated.

Kramers suggested that the orbits of a quantum system should be Fourier analyzed, decomposed into harmonics at multiples of the orbit frequency:

$$X_n(t) = \sum_{k=-\infty}^{\infty} e^{ik\omega t} X_{n;k}$$

The index n describes the quantum numbers of the orbit, it would be n–l–m in the Sommerfeld model. The frequency ω is the angular frequency of the orbit 2π/Tn while k is an index for the Fourier mode. Bohr had suggested that the k-th harmonic of the classical motion correspond to the transition from level n to level n−k.

Kramers proposed that the transition between states were analogous to classical emission of radiation, which happens at frequencies at multiples of the orbit frequencies. The rate of emission of radiation is proportional to $|X_k|^2$,, as it would be in classical mechanics. The description was approximate, since the Fourier components did not have frequencies that exactly match the energy spacings between levels.

Limitations of the Old Quantum Theory

The old quantum theory had some limitations:

- The old quantum theory provides no means to calculate the intensities of the spectral lines.
- It fails to explain the anomalous Zeeman effect (that is, where the spin of the electron cannot be neglected).
- It cannot quantize “chaotic” systems, i.e. dynamical systems in which trajectories are neither closed nor periodic and whose analytical form does not exist. This presents a problem for systems as simple as a 2-electron atom which is classically chaotic analogously to the famous gravitational three-body problem.

However it can be used to describe atoms with more than one electron (e.g. Helium) and the Zeeman effect. It was later proposed that the old quantum theory is in fact the semi-classical approximation to the canonical quantum mechanics but its limitations are still under investigation.

The old quantum theory was sparked by the 1900 work of Max Planck on the emission and absorption of light, and began in earnest after the work of Albert Einstein on the specific heats of solids. Einstein, followed by Debye, applied quantum principles to the motion of atoms, explaining the specific heat anomaly.

In 1913, Niels Bohr identified the correspondence principle and used it to formulate a model of the hydrogen atom which explained the line spectrum. In the next few years Arnold Sommerfeld extended the quantum rule to arbitrary integrable systems making use of the principle of adiabatic invariance of the quantum numbers introduced by Lorentz and Einstein. Sommerfeld›s model was much closer to the modern quantum mechanical picture than Bohr›s.

Throughout the 1910s and well into the 1920s, many problems were attacked using the old quantum theory with mixed results. Molecular rotation and vibration spectra were understood and the electron›s spin was discovered, leading to the confusion of half-integer quantum numbers. Max Planck introduced the zero point energy and Arnold Sommerfeld semiclassically quantized the relativistic hydrogen atom. Hendrik Kramers explained the Stark effect. Bose and Einstein gave the correct quantum statistics for photons.

Kramers gave a prescription for calculating transition probabilities between quantum states in terms of Fourier components of the motion, ideas which were extended in collaboration with Werner Heisenberg to a semiclassical matrix-like description of atomic transition probabilities. Heisenberg went on to reformulate all of quantum theory in terms of a version of these transition matrices, creating matrix mechanics.

In 1924, Louis de Broglie introduced the wave theory of matter, which was extended to a semiclassical equation for matter waves by Albert Einstein a short time later. In 1926 Erwin Schrödinger found a completely quantum mechanical wave-equation, which reproduced all the successes of the old quantum theory without ambiguities and inconsistencies. Schrödinger›s wave mechanics developed separately from matrix mechanics until Schrödinger and others proved that the two methods predicted the same experimental consequences. Paul Dirac later proved in 1926 that both methods can be obtained from a more general method called transformation theory.

REFERENCES

1. ter Haar, D. (1967). The Old Quantum Theory. Pergamon Press. p. 206. ISBN 0-08-012101-2.
2. Sommerfeld, Arnold (1919). Atombau und Spektrallinien’. Braunschweig: Friedrich Vieweg und Sohn. ISBN 3-87144-484-7.
3. Arnold Sommerfeld (1921). Atombau und Spektrallinien. Braunschweig.
4. Ya I Granovski (2004). “Sommerfeld formula and Dirac’s theory” (PDF). Physics-Uspekhi. 47 (5): 523–524. Bibcode:2004PhyU...47..523G. doi:10.1070/PU2004v047n05ABEH001885.
5. Einstein, Albert (1905). “Über einen die Erzeugung und Verwandlung des Lichtes betreffenden heuristischen Gesichtspunkt [On a Heuristic Point of View Concerning the Production and Transformation of Light] (PDF). Annalen der Physik (in German). 17 (6): 132–148. Bibcode:1905AnP...322..132E. doi:10.1002/andp.19053220607. Retrieved 2008-02-18.
6. Chaddha, G.S. (2006). Quantum Mechanics. New Dehli: New Age international. pp. 8–9. ISBN 81-224-1465-6.
7. E.A. Solov’ev, E. A. (2011). Classical approach in atomic physics. European Physical Journal D. 65 (3): 331–351. arXiv:1003.4387 . Bibcode:2011EPJD...65..331S. doi:10.1140/epjd/e2011-20261-6.
8. L.D. Landau, E.M. Lifshitz (1977). Quantum Mechanics: Non-Relativistic Theory. Vol. 3 (3rd ed.). Pergamon Press. ISBN 978-0-08-020940-1.

CHAPTER 3

Fundamentals of Quantum Mechanics

Chapter Contents

- **Introduction**
- **Quantum Decoherence**
- **Examples of Non-Unitary Modelling of Decoherence**
- **Mathematical Details**
- **Quantum Entanglement**
- **Energy Level**
- **Measurement in Quantum Mechanics**

INTRODUCTION

In physics, complementarity is both a theoretical and an experimental result of quantum mechanics, also referred to as principle of complementarity, closely associated with the Copenhagen interpretation. It holds that objects have complementary properties which cannot be observed or measured at all at the same time.

The complementarity principle was formulated by Niels Bohr, a leading founder of quantum mechanics.Examples of complementary properties that Bohr considered:

- Position and momentum
- Energy and duration
- Spin on different axes
- Wave and particle
- Value of a field and its change (at a certain position)
- Entanglement and coherence

Bohr's complementary principle should not be confused with observables that do not commute, like position and momentum. This is an entirely different subject.

Concept

For example, the particle and wave aspects of physical objects are such complementary phenomena. Both concepts are borrowed from classical mechanics, where it is impossible to be a particle and wave at the same time. Therefore, it is impossible to measure the full properties of the wave and particle at a particular moment. Moreover, Bohr implies that it is not possible to regard objects governed by quantum mechanics as having intrinsic properties independent of determination with a measuring device. The type of measurement determines which property is shown. However the single and double-slit experiment and other experiments show that some effects of wave and particle can be measured in one measurement.

Nature

An aspect of complementarity is that it not only applies to measurability or knowability of some property of a physical entity, but more importantly it applies to the limitations of that physical entity's very manifestation of the property in the physical world. All properties of physical entities exist only in pairs, which Bohr described as complementary or conjugate pairs. Physical reality is determined and defined by manifestations of properties which are limited by trade-offs between these complementary pairs. For example, an electron can manifest a greater and greater accuracy of its position only in even trade for a complementary loss in accuracy of manifesting its momentum. This means that there is a limitation on the precision with which an electron can possess (i.e., manifest) position, since an infinitely precise position would dictate that its manifested momentum would be infinitely imprecise, or undefined (i.e., non-manifest or not possessed), which is not possible. The ultimate limitations in precision of property manifestations are quantified by the Heisenberg uncertainty principle and Planck units. Complementarity and Uncertainty dictate that therefore all properties and actions in the physical world manifest themselves as non-deterministic to some degree.

Physicists F.A.M. Frescura and Basil Hiley have summarized the reasons for the introduction of the principle of complementarity in physics as follows:

"In the traditional view, it is assumed that there exists a reality in space-time and that this reality is a given thing, all of whose aspects can be viewed or articulated at any given moment. Bohr was the first to point out that quantum mechanics called this traditional outlook into question. To him the 'indivisibility of the quantum of action', which was his way of describing the uncertainty principle, implied that not all aspects of a system can be viewed simultaneously. By using one particular piece of apparatus only certain features could be made manifest at the expense of others, while with a different piece of apparatus another complementary aspect could be made manifest in such a way that the original set became non-manifest, that is, the original attributes were no longer well defined. For Bohr, this was an indication that the principle of complementarity, a principle that he had previously known to appear extensively in other intellectual disciplines but which did not appear in classical physics, should be adopted as a universal principle." The emergence of complementarity in a system occurs when one considers the circumstances under which one attempts to measure its properties; as Bohr noted, the principle of complementarity "implies the impossibility of any sharp separation between the behaviour of atomic objects and the interaction with the measuring instruments that serve to define the conditions under which the phenomena appear." It is important to distinguish, as did Bohr in his original statements, the principle of complementarity from a statement of the uncertainty principle. For a technical discussion of contemporary issues surrounding complementarity in physics see, e.g., Bandyopadhyay (2000), from which parts of this discussion were drawn.

Additional Considerations

In his original lecture on the topic, Bohr pointed out that just as the finitude of the speed of light implies the impossibility of a sharp separation between space and time (relativity), the finitude of the quantum of action implies the impossibility of a sharp separation between the behavior of a system and its interaction with the measuring instruments and leads to the well known difficulties with the concept of 'state' in quantum theory; the notion of complementarity is intended to symbolize this new situation in epistemology created by quantum theory. Some people consider it a philosophical adjunct to quantum mechanics, while others consider it to be a discovery that is as important as the formal aspects of quantum theory. Examples of the latter include Leon Rosenfeld, who claimed that omplementarity is not a philosophical superstructure invented by Bohr to be placed as a decoration on top of the quantal formalism, it is the bedrock of the quantal description.", and John Wheeler, who opined that "Bohr's principle of complementarity is the most revolutionary scientific concept of this century and the heart of his fifty-year search for the full significance of the quantum idea."

Experiments

The quintessential example of wave–particle complementarity in the laboratory is the double slit. The crux of the complementary behavior is the question: "What information exists – embedded in the constituents of the universe – that can reveal the history of the signal particles as they pass through the double slit?" If information exists (even if it is not measured by a conscious observer) that reveals "which slit" each particle traversed, then each particle will exhibit no wave interference with the other slit. This is the particle-like behavior. But if no information exists about which slit – so that no conscious observer, no matter how well equipped, will ever be able to determine which slit each particle traverses – then the signal particles will interfere with themselves as if they traveled through both slits at the same time, as a wave. This is the wave-like behavior. These behaviors are complementary, according to the Englert–Greenberger duality relation, because when one behavior is observed the other is absent. Both behaviors can be observed at the same time, but each only as lesser manifestations of their full behavior (as determined by the duality relation). This superposition of complementary behaviors exists whenever there is partial "which slit" information. While there is some contention to the duality relation, and thus complementarity itself, the contrary position is not accepted by mainstream physics. Double slit experiments with single photons show clearly that photons are particles at the same time as they are waves. The Photons impact the screen where they are detected in points and when enough points have accumulated the wave aspect is clearly visible. Also the particle and wave aspect is seen at the same time in photons that are stationary.

Various neutron interferometry experiments demonstrate the subtlety of the notions of duality and complementarity. By passing through the interferometer, the neutron appears to act as a wave. Yet upon passage, the neutron is subject to gravitation. As the neutron interferometer is rotated through Earth's gravitational field a phase change between the two arms of the interferometer can be observed, accompanied by a change in the constructive and destructive interference of the neutron waves on exit from the interferometer. Some interpretations claim that understanding the interference effect requires one to concede that a single neutron takes both paths through the interferometer at the same time; a single neutron would "be in two places at once", as it were. Since the two paths through a neutron interferometer can be as far as 5 cm to 15 cm apart, the effect is hardly microscopic. This is similar to traditional double-slit and mirror interferometer experiments where the slits (or mirrors) can be arbitrarily far apart. So, in interference and diffraction experiments, neutrons behave the same way as photons (or electrons) of corresponding wavelength.

Niels Bohr apparently conceived of the principle of complementarity during a skiing vacation in Norway in February and March 1927, during which he received a letter from Werner Heisenberg regarding the latter's newly discovered (and not yet published) uncertainty principle. Upon returning from his vacation, by which time Heisenberg had already submitted his paper on the uncertainty principle for publication, he convinced Heisenberg that the uncertainty principle was a manifestation of the deeper concept of complementarity. Heisenberg duly appended a note to this effect to his paper on the uncertainty principle, before its publication, stating:

Bohr has brought to my attention that the uncertainty in our observation does not arise exclusively from the occurrence of discontinuities, but is tied directly to the demand that we ascribe equal validity to the quite different experiments which show up in the particulate theory on one hand, and in the wave theory on the other hand.

Bohr publicly introduced the principle of complementarity in a lecture he delivered on 16 September 1927 at the International Physics Congress held in Como, Italy, attended by most of the leading physicists of the era, with the notable exceptions of Einstein, Schrödinger, and Dirac. However, these three were in attendance one month later when Bohr again presented the principle at the Fifth Solvay Congress in Brussels, Belgium. The lecture was published in the proceedings of both of these conferences, and was republished the following year in Naturwissenschaften (in German) and in Nature (in English).

An article written by Bohr in 1949 titled "Discussions with Einstein on Epistemological Problems in Atomic Physics" is considered by many to be a definitive description of the notion of complementarity.

QUANTUM DECOHERENCE

Quantum decoherence is the loss of quantum coherence. In quantum mechanics, particles such as electrons behave like waves and are described by a wavefunction. These waves can interfere, leading to the peculiar behaviour of quantum particles. As long as there exists a definite phase relation between different states, the system is said to be coherent. This coherence is a fundamental property of quantum mechanics, and is necessary for the function of quantum computers. However, when a quantum system is not perfectly isolated, but in contact with its surroundings, the coherence decays with time, a process called quantum decoherence. As a result of this process, the quantum behaviour is lost. Decoherence was first introduced in 1970 by the German physicist H. Dieter Zeh and has been a subject of active research since the 1980s.

Decoherence can be viewed as the loss of information from a system into the environment (often modeled as a heat bath), since every system is loosely coupled with the energetic state of its surroundings. Viewed in isolation, the system's dynamics are non-unitary (although the combined system plus environment evolves in a unitary fashion). Thus the dynamics of the system alone are irreversible. As with any coupling, entanglements are generated between the system and environment. These have the effect of sharing quantum information with—or transferring it to—the surroundings.

Decoherence does not generate actual wave function collapse. It only provides an explanation for the observation of wave function collapse, as the quantum nature of the system "leaks" into the environment. That is, components of the wavefunction are decoupled from a coherent system, and acquire phases from their immediate surroundings. A total superposition of the global or universal wavefunction still exists (and remains coherent at the global level), but its ultimate fate remains an interpretational issue. Specifically, decoherence does not attempt to explain the measurement problem. Rather, decoherence provides an explanation for the transition of the system to a mixture of states that seem to correspond to those states observers perceive. Moreover, our observation tells us that this mixture looks like a proper quantum ensemble in a measurement situation, as we observe that measurements lead to the "realization" of precisely one state in the "ensemble".

Decoherence represents a challenge for the practical realization of quantum computers, since such machines are expected to rely heavily on the undisturbed evolution of quantum coherences. Simply put, they require that coherent states be preserved and that decoherence is managed, in order to actually perform quantum computation.

Mechanisms

To examine how decoherence operates, an "intuitive" model is presented. The model requires some familiarity with quantum theory basics. Analogies are made between visualisable classical phase spaces and Hilbert spaces. A more rigorous derivation in Dirac notation shows how decoherence destroys interference effects and the "quantum nature" of systems. Next, the density matrix approach is presented for perspective.

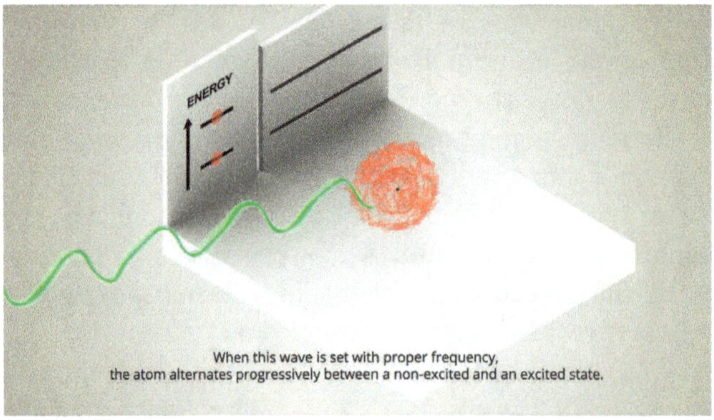

Quantum superposition of states and decoherence measurement through Rabi oscillations

Phase Space Picture

An N-particle system can be represented in non-relativistic quantum mechanics by a wavefunction, $\psi(x_1, x_2, \ldots, x_N)$ where each x_i is a point in 3-dimensional space. This has analogies with the classical phase space. A classical phase space contains a real-valued function in 6N dimensions (each particle contributes 3 spatial coordinates and 3 momenta). Our "quantum" phase space, on the other hand, involves a complex-valued function on a 3N dimensional space. The position and momenta are represented by operators that do not commute, and ψ lives in the mathematical structure of a Hilbert space. Aside from these differences, however, the rough analogy holds.

Different previously-isolated, non-interacting systems occupy different phase spaces. Alternatively we can say they occupy different, lower-dimensional subspaces in the phase space of the joint system. The effective dimensionality of a system's phase space is the number of degrees of freedom present which—in non-relativistic models—is 6 times the number of a system's free particles. For a macroscopic system this will be a very large dimensionality. When two systems (and the environment would be a system) start to interact, though, their associated state vectors are no longer constrained to the subspaces. Instead the combined state vector time-evolves a path through the "larger volume", whose dimensionality is the sum of the dimensions of the two subspaces. The extent to which two vectors interfere with each other is a measure of how "close" they are to each other (formally, their overlap or Hilbert space multiplies together) in the phase space. When a system couples to an external environment, the dimensionality of, and hence "volume" available to, the joint state vector increases enormously. Each environmental degree of freedom contributes an extra dimension.

The original system's wavefunction can be expanded in many different ways as a sum of elements in a quantum superposition. Each expansion corresponds to a projection of the wave vector onto a basis. The basis can be chosen at will. Let us choose an expansion where the resulting basis elements interact with the environment in an element-specific way. Such elements will—with overwhelming probability—be rapidly separated from each other by their natural unitary time evolution along their own independent paths. After a very short interaction, there is almost no chance of any further interference. The process is effectively irreversible. The different elements effectively become "lost" from each other in the expanded phase space created by coupling with the environment; in phase space, this decoupling is monitored through the Wigner quasi-probability distribution. The original elements are said to have decohered. The environment has effectively selected out those expansions or decompositions of the original state vector that decohere (or lose phase coherence) with each other. This is called "environmentally-induced-superselection", or einselection. The decohered elements of the system no longer exhibit quantum interference between each other, as in a double-slit experiment. Any elements that decohere from each other via environmental interactions are said to be quantum entangled with the environment. The converse is not true: not all entangled states are decohered from each other.

Any measuring device or apparatus acts as an environment since, at some stage along the measuring chain, it has to be large enough to be read by humans. It must possess a very large number of hidden degrees of freedom. In effect, the interactions may be considered to be quantum measurements. As a result of an interaction, the wave functions of the system and the measuring device become entangled with each other. Decoherence happens when different portions of the system's wavefunction become entangled in different ways with the measuring device. For two einselected elements of the entangled system's state to interfere, both the original system and the measuring in both elements device must significantly overlap, in the scalar product sense. If the measuring device has many degrees of freedom, it is very unlikely for this to happen.

As a consequence, the system behaves as a classical statistical ensemble of the different elements rather than as a single coherent quantum superposition of them. From the perspective of each ensemble member's measuring device, the system appears to have irreversibly collapsed onto a state with a precise value for the measured attributes, relative to that element.

Dirac Notation

Using Dirac notation, let the system initially be in the state

$$|\psi\rangle = \sum_i |i\rangle\langle i|\psi\rangle$$

where the $|i\rangle$ s form an einselected basis (environmentally induced selected eigenbasis), and let the environment initially be in the state $|\epsilon\rangle$. The vector basis of the combination of the system and the environment consists of the tensor products of the basis vectors of the two subsystems. Thus, before any interaction between the two subsystems, the joint state can be written as

$$|\text{before}\rangle = \sum_i |i\rangle|\epsilon\rangle\langle i|\psi\rangle,$$

where $|i\rangle|\epsilon\rangle$ is shorthand for the tensor product: $|i\rangle \otimes |\epsilon\rangle$. There are two extremes in the way the system can interact with its environment: either (1) the system loses its distinct identity and merges with the environment (e.g. photons in a cold, dark cavity get converted into molecular excitations within the cavity walls), or (2) the system is not disturbed at all, even though the environment is disturbed (e.g. the idealized non-disturbing measurement). In general an interaction is a mixture of these two extremes that we examine.

System Absorbed by Environment

If the environment absorbs the system, each element of the total system's basis interacts with the environment such that

$|i\rangle|\epsilon\rangle$ evolves into $|\epsilon_i\rangle$

and so

$|\textbf{before}\rangle$ evolves into $|\textbf{after}\rangle = \sum_i |\epsilon_i\rangle\langle i|\psi\rangle.$

The unitarity of time-evolution demands that the total state basis remains orthonormal, i.e. the scalar or inner products of the basis vectors must vanish, since $\langle i|j\rangle = \delta_{ij}$:

$$\langle\epsilon_i|\epsilon_j\rangle = \delta_{ij}.$$

This orthonormality of the environment states is the defining characteristic required for einselection.

System Not Disturbed by Environment

In an idealised measurement, the system disturbs the environment, but is itself undisturbed by the environment. In this case, each element of the basis interacts with the environment such that

$|i\rangle|\epsilon\rangle$ evolves into the product $|i,\epsilon_i\rangle = |i\rangle|\epsilon_i\rangle$

and so

$|\textbf{before}\rangle$ evolves into $|\textbf{after}\rangle = \sum_i |i,\epsilon_i\rangle\langle i|\psi\rangle.$

In this case, unitarity demands that

$$\langle i,\epsilon_i|j,\epsilon_j\rangle = \langle i|j\rangle\langle\epsilon_i|\epsilon_j\rangle = \delta_{ij}\langle\epsilon_i|\epsilon_j\rangle = \delta_{ij}.$$

Additionally, decoherence requires, by virtue of the large number of hidden degrees of freedom in the environment, that

$$\langle\epsilon_i|\epsilon_j\rangle \approx \delta_{ij}.$$

As before, this is the defining characteristic for decoherence to become einselection. The approximation becomes more exact as the number of environmental degrees of freedom affected increases.

Note that if the system basis $|i\rangle$ were not an einselected basis then the last condition is trivial since the disturbed environment is not a function of i and we have the trivial disturbed environment basis $|\epsilon_j\rangle = |\epsilon'\rangle$ This would correspond to the system basis being degenerate with respect to the environmentally-defined-measurement-observable. For a complex environmental interaction (which would be expected for a typical macroscale interaction) a non-einselected basis would be hard to define.

Loss of Interference and the Transition from Quantum to Classical

The utility of decoherence lies in its application to the analysis of probabilities, before and after environmental interaction, and in particular to the vanishing of quantum interference terms after decoherence has occurred. If we ask what is the probability of observing the system making a transition from ψ to ϕ before ψ has interacted with its environment, then application of the Born probability rule states that the transition probability is the modulus squared of the scalar product of the two states:

$$prob_{before}(\psi \to \phi) = |\langle\psi|\phi\rangle|^2 = |\sum_i \psi_i^* \phi_i|^2 = \sum_i |\psi_i^* \phi_i|^2 + \sum_{ij;i\neq j} \psi_i^* \psi_j \phi_j^* \phi_i$$

where $\psi_i = \langle i|\psi\rangle, \psi_i^* = \langle\psi|i\rangle$ and $\phi_i = \langle i|\phi\rangle$ etc.

Terms appear in the expansion of the transition probability above which involve $i \neq j$; these can be thought of as representing interference between the different basis elements or quantum alternatives. This is a purely quantum effect and represents the non-additivity of the probabilities of quantum alternatives.

To calculate the probability of observing the system making a quantum leap from ψ to ϕ before ψ has interacted with its environment, then application of the Born probability rule states we must sum over all the relevant possible states of the environment, E_i before squaring the modulus:

$$prob_{after}(\psi \to \phi) = \sum_j |\langle after|\phi, \epsilon_j\rangle|^2 = \sum_j |\sum_i \psi_i^* \langle i, \epsilon_i|\phi, \epsilon_j\rangle|^2 = \sum_j |\sum_i \psi_i^* \langle i|\phi\rangle \langle \epsilon_i|\epsilon_j\rangle|^2$$

The internal summation vanishes when we apply the decoherence / einselection condition $\langle \epsilon_i|\epsilon_j\rangle \approx \delta_{ij}$

$$prob_{after}(\psi \to \phi) \approx \sum_j |\psi_j^* \langle j|\phi\rangle|^2 = \sum_i |\psi_i^* \phi_i|^2$$

If we compare this with the formula we derived before the environment introduced decoherence we can see that the effect of decoherence has been to move the summation sign Σ_i from inside of the modulus sign to outside. As a result, all the cross- or quantum interference-terms:

$$\sum_{ij;i\neq j} \psi_i^* \psi_j \phi_j^* \phi_i$$

have vanished from the transition probability calculation. The decoherence has irreversibly converted quantum behaviour (additive probability amplitudes) to classical behaviour (additive probabilities).

In terms of density matrices, the loss of interference effects corresponds to the diagonalization of the "environmentally traced over" density matrix.

Density Matrix Approach

The effect of decoherence on density matrices is essentially the decay or rapid vanishing of the off-diagonal elements of the partial trace of the joint system's density matrix, i.e. the trace, with respect to any environmental basis, of the density matrix of the combined system and its environment. The decoherence irreversibly converts the "averaged" or "environmentally traced over" density matrix from a pure state to a reduced mixture; it is this that gives the appearance of wavefunction collapse. Again this is called "environmentally-induced-superselection", or einselection. The advantage of taking the partial trace is that this procedure is indifferent to the environmental basis chosen.

Initially, the density matrix of the combined system can be denoted as,

$$\rho = |before\rangle\langle before| = |\psi\rangle\langle\psi| \otimes |\epsilon\rangle\langle\epsilon|$$

Where $|\epsilon\rangle$ is the state of the environment. Then if the transition happens before any interaction takes place between the system and the environment, the environment subsystem has no part and can be traced out, leaving the reduced density matrix for the system,

$$\rho_{sys} = Tr_{env}(\rho) = |\psi\rangle\langle\psi|\langle\epsilon|\epsilon\rangle = |\psi\rangle\langle\psi|$$

Now the transition probability will be given as:

$$prob_{before}(\psi \to \phi) = \langle\phi|\rho_{sys}|\phi\rangle = \langle\phi|\psi\rangle\langle\psi|\phi\rangle = |\langle\psi|\phi\rangle|^2 = \sum_i |\psi_i^*\phi_i|^2 + \sum_{ij;i\neq j} \psi_i^*\psi_j\phi_j^*\phi_i$$

where $\psi_i = \langle i|\psi\rangle, \psi_i^* = \langle\psi|i\rangle$ and $\phi_i = \langle i|\phi\rangle$ etc.

Now the case when transition takes place after the interaction of the system with the environment. The combined density matrix will be,

$$\rho = |after\rangle\langle after| = \sum_{i,j} \psi_i\psi_j^* |i, \epsilon_i\rangle\langle j, \epsilon_j| = \sum_{i,j} \psi_i\psi_j^* |i\rangle\langle j| \otimes |\epsilon_i\rangle\langle\epsilon_j|$$

To get the reduced density matrix of the system we trace out the environment and employ the decoherence/einselection condition and see that the off-diagonal terms vanish,

$$\rho_{sys} = Tr_{env}(\sum_{i,j} \psi_i\psi_j^* |i\rangle\langle j| \otimes |\epsilon_i\rangle\langle\epsilon_j|) = \sum_{i,j} \psi_i\psi_j^* |i\rangle\langle j|\langle\epsilon_i|\epsilon_j\rangle = \sum_{i,j} \psi_i\psi_j^* |i\rangle\langle j|\delta_{ij} = \sum_i |\psi_i|^2 |i\rangle\langle i|$$

Similarly the final reduced density matrix after the transition will be $\sum_j |\phi_j|^2 |j\rangle\langle j|$ The transition probability will then be given as:

$$prob_{after}(\psi \to \phi) = \sum_{i,j} |\psi_i|^2 |\phi_j|^2 \langle j|i\rangle\langle i|j\rangle = \sum_i |\psi_i^*\phi_i|^2$$

which has no contribution from the interference terms, $\sum_{ij;i\neq j} \psi_i^* \psi_j \phi_j^* \phi_i$.

The density matrix approach has been combined with the Bohmian approach to yield a reduced trajectory approach, taking into account the system reduced density matrix and the influence of the environment.[7]

Operator-Sum Representation

Consider a system S and environment (bath) B, which are closed and can be treated quantum mechanically. Let $\mathcal{H}_S$ and $\mathcal{H}_B$ be the system's and bath's Hilbert spaces, respectively. Then the Hamiltonian for the combined system is

$$\hat{H} = \hat{H}_S \otimes \hat{I}_B + \hat{I}_S \otimes \hat{H}_B + \hat{H}_I$$

where $\hat{H}_S, \hat{H}_B$ are the system and bath Hamiltonians, respectively, and $\hat{H}_I$ is the interaction Hamiltonian between the system and bath, and $\hat{I}_S, \hat{I}_B$ are the identity operators on the system and bath Hilbert spaces, respectively. The time-evolution of the density operator of this closed system is unitary and, as such, is given by

$$\rho_{SB}(t) = \hat{U}(t)\rho_{SB}(0)\hat{U}^\dagger(t)$$

where the unitary operator is $\hat{U} = e^{\frac{-i\hat{H}t}{\hbar}}$ If the system and bath are not entangled initially, then we can write $\rho_{SB} = \rho_S \otimes \rho_B$. Therefore, the evolution of the system becomes

$$\rho_{SB}(t) = \hat{U}(t)[\rho_S(0) \otimes \rho_B(0)]\hat{U}^\dagger(t).$$

where $\hat{S}_i \otimes \hat{B}_i$ is the operator acting on the combined system-bath Hilbert space, and $\hat{S}_i, \hat{B}_i$ are the operators that act on the system and bath, respectively. This coupling of the system and bath is the cause of decoherence in the system alone. To see this, a partial trace is performed over the bath to give a description of the system alone:

$$\rho_S(t) = Tr_B[\hat{U}(t)[\rho_S(0) \otimes \rho_B(0)]\hat{U}^\dagger(t)].$$

$\rho_S(t)$ is called the reduced density matrix and gives information about the system only. If the bath is written in terms of its set of orthogonal basis kets, that is, if it has been initially diagonalized then $\rho_B(0) = \sum_j a_j |j\rangle\langle j|$. Computing the partial trace with respect to this (computational) basis gives:

$$\rho_S(t) = \sum_l \hat{A}_l \rho_S(0) \hat{A}_l^\dagger$$

where $\hat{A}_l, \hat{A}_l^\dagger$ are defined as the **Kraus operators** and are represented as

$$\hat{A}_l = \sqrt{a_j}\langle k|\hat{U}|j\rangle.$$

This is known as the **operator-sum representation** (OSR). A condition on the Kraus operators can be obtained by using the fact that $Tr(\rho_S(t)) = 1$; this then gives

$$\sum_l \hat{A}_l^\dagger \hat{A}_l = \hat{I}_S.$$

This restriction determines if decoherence will occur or not in the OSR. In particular, when there is more than one term present in the sum for $\rho_S(t)$ then the dynamics of the system will be non-unitary and hence decoherence will take place.

Semigroup Approach

A more general consideration for the existence of decoherence in a quantum system is given by the **master equation**, which determines how the density matrix of the system alone evolves in time (see also the Belavkin equation for the evolution under continuous measurement). This uses the Schrödinger picture, where evolution of the state (represented by its density matrix) is considered. The master equation is:

$$\rho'_S(t) = \frac{-i}{\hbar}\left[\tilde{\mathbf{H}}_S, \rho_S(t)\right] + L_D\left[\rho_S(t)\right]$$

where $\tilde{\mathbf{H}}_S = \mathbf{H}_S + \Delta$ is the system Hamiltonian, $\mathbf{H}_S$, along with a (possible) unitary contribution from the bath, Δ and L_D is the **Lindblad decohering term**.[3] The Lindblad decohering term is represented as

$$L_D\left[\rho_S(t)\right] = \frac{1}{2}\sum_{\alpha,\beta=1}^{M} b_{\alpha\beta}\left(\left[\mathbf{F}_\alpha, \rho_S(t)\mathbf{F}_\beta^\dagger\right] + \left[\mathbf{F}_\alpha\rho_S(t), \mathbf{F}_\beta^\dagger\right]\right).$$

The $\{\mathbf{F}_\alpha\}_{\alpha=1}^{M}$ are basis operators for the M-dimensional space of bounded operators that act on the system Hilbert space $\mathcal{H}_S$-these are the **error generators**-and $b_{\alpha\beta}$ represent the elements of a positive semi-definite Hermitian matrix-these matrix elements characterize the decohering processes and, as such, are called the **noise parameters**. The semigroup approach is particularly nice, because it distinguishes between the unitary and decohering(non-unitary) processes, which is not the case with the OSR. In particular, the non-unitary dynamics are represented by L_D whereas the unitary dynamics of the state are represented by the usual Heisenberg commutator. Note that when $L_D\left[\rho_S(t)\right] = 0$, the dynamical evolution of the system is unitary. The conditions for the evolution of the system density matrix to be described by the master equation are:

(1) the evolution of the system density matrix is determined by a one-parameter semigroup
(2) the evolution is "completely positive" (i.e. probabilities are preserved)
(3) the system and bath density matrices are initially decoupled.

EXAMPLES OF NON-UNITARY MODELLING OF DECOHERENCE

Decoherence can be modelled as a non-unitary process by which a system couples with its environment (although the combined system plus environment evolves in a unitary fashion). Thus the dynamics of the system alone, treated in isolation, are non-unitary and, as such, are represented by irreversible transformations acting on the system's Hilbert space,H Since the system's dynamics are represented by irreversible representations, then any information present in the quantum system can be lost to the environment or heat bath. Alternatively, the decay of quantum information caused by the coupling of the system to the environment is referred to as decoherence. Thus decoherence is the process by which information of a quantum system is altered by the system's interaction with its environment (which form a closed system), hence creating an entanglement between the system and heat bath (environment). As such, since the system is entangled with its environment in some unknown way, a description of the system by itself cannot be made without also referring to the environment (i.e. without also describing the state of the environment).

Rotational Decoherence

Consider a system of N qubits that is coupled to a bath symmetrically. Suppose this system of N qubits undergoes a rotation around the $|\uparrow\rangle\langle\uparrow|, |\downarrow\rangle\langle\downarrow|$ $(|0\rangle\langle 0|, |1\rangle\langle 1|)$ eigenstates of $\hat{J}_z$. Then under such a rotation, a random phase, ϕ, will be created between the eigenstates $|0\rangle, |1\rangle$ of $\hat{J}_z$. Thus these basis qubits $|0\rangle$ and $|1\rangle$ will transform in the following way:

$$|0\rangle \to |0\rangle, |1\rangle \to e^{i\phi}|1\rangle.$$

This transformation is performed by the rotation operator

$$R_z(\phi) = \begin{pmatrix} 1 & 0 \\ 0 & e^{i\phi} \end{pmatrix}.$$

Since any qubit in this space can be expressed in terms of the basis qubits, then all such qubits will be transformed under this rotation. Consider a qubit in a pure state $|\psi_j\rangle = a|0_j\rangle + b|1_j\rangle$. This state will decohere since it is not "encoded" with the dephasing factor $e^{i\phi}$ This can be seen by examining the density matrix averaged over all values of ϕ:

$$\rho_j = \int_{-\infty}^{\infty} R_z(\phi)|\psi_j\rangle\langle\psi_j|R_z^\dagger(\phi)p(\phi)\,d\phi$$

Where $p(\phi)$ is a probability density. If $p(\phi)$ is given as a Gaussian distribution

$$p(\phi) = (4\pi\alpha)^{-\frac{1}{2}} e^{\frac{-\phi^2}{4\alpha}}$$

then the density matrix is

$$\rho_j = \begin{pmatrix} |a|^2 & ab^* e^{-\alpha} \\ a^* b e^{-\alpha} & |b|^2 \end{pmatrix}.$$

Since the off-diagonal elements—the coherence terms—decay for increasing α, then the density matrices for the various qubits of the system will be indistinguishable. This means that no measurement can distinguish between the qubits, thus creating decoherence between the various qubit states. In particular, this dephasing process causes the qubits to collapse onto the $|0\rangle\langle 0|, |1\rangle\langle 1|$ axis. This is why this type of decoherence process is called **collective dephasing**, because the mutual phases between all qubits of the N-qubit system are destroyed.

Depolarizing

Depolarizing is a non-unitary transformation on a quantum system which maps pure states to mixed states. This is a non-unitary process, because any transformation that reverses this process will map states out of their respective Hilbert space thus not preserving positivity (i.e. the original probabilities are mapped to negative probabilities, which is not allowed). The 2-dimensional case of such a transformation would consist of mapping pure states on the surface of the Bloch sphere to mixed states within the Bloch sphere. This would contract the Bloch sphere by some finite amount and the reverse process would expand the Bloch sphere, which cannot happen.

Dissipation

Dissipation is a decohering process by which the populations of quantum states are changed due to entanglement with a bath. An example of this would be a quantum system that can exchange its energy with a bath through the interaction Hamiltonian. If the system is not in its ground state and the bath is at a temperature lower than that of the system's, then the system will give off energy to the bath and thus higher-energy eigenstates of the system Hamiltonian will decohere to the ground state after cooling and, as such, they will all be non-degenerate. Since the states are no longer degenerate, then they are not distinguishable and thus this process is irreversible (non-unitary).

Timescales

Decoherence represents an extremely fast process for macroscopic objects, since these are interacting with many microscopic objects, with an enormous number of degrees of freedom, in their natural environment. The process explains why we tend not to observe quantum behaviour in everyday macroscopic objects. It also explains why we do see classical fields emerge from the properties of the interaction between matter and radiation for large amounts of matter. The time taken for off-diagonal components of the density matrix to effectively vanish is called the **decoherence time**, and is typically extremely short for everyday, macroscale processes.

Measurement

The discontinuous "wave function collapse" postulated in the Copenhagen interpretation to enable the theory to be related to the results of laboratory measurements now can be understood as an aspect of the normal dynamics of quantum mechanics via the decoherence process. Consequently, decoherence is an important part of the modern alternative to the Copenhagen interpretation, based on consistent histories. Decoherence shows how a macroscopic system interacting with a lot of microscopic systems (e.g. collisions with air molecules or photons) moves from being in a pure quantum state—which in general will be a coherent superposition (see Schrödinger's cat)—to being in an incoherent mixture of these states. The weighting of each outcome in the mixture in case of measurement is exactly that which gives the probabilities of the different results of such a measurement.

However, decoherence by itself may not give a complete solution of the measurement problem, since all components of the wave function still exist in a global superposition, which is explicitly acknowledged in the many-worlds interpretation. All decoherence explains, in this view, is why these coherences are no longer available for inspection by local observers. To present a solution to the measurement problem in most interpretations of quantum mechanics, decoherence must be supplied with some nontrivial interpretational considerations (as for example Wojciech Zurek tends to do in his Existential interpretation). However, according to Everett and DeWitt the many-worlds interpretation can be derived from the formalism alone, in which case no extra interpretational layer is required.

MATHEMATICAL DETAILS

We assume for the moment the system in question consists of a subsystem being studied, A and the "environment" ϵ, and the total Hilbert space is the tensor product of a Hilbert space describing A, H_A and a Hilbert space describing ϵ, H_ϵ that is,

$$H = H_A \otimes H_\epsilon.$$

This is a reasonably good approximation in the case where A and ϵ, are relatively independent (e.g. there is nothing like parts of A mixing with parts of ϵ, or vice versa). The point is, the interaction with the environment is for all practical purposes unavoidable (e.g. even a single excited atom in a vacuum would emit a photon which would then go off). Let's say this interaction is described by a unitary transformation U acting upon H. Assume the initial state of the environment is $|\mathrm{in}\rangle$ and the initial state of A is the superposition state

$$c_1|\psi_1\rangle + c_2|\psi_2\rangle$$

Where $|\psi_1\rangle$ and $|\psi_2\rangle$ are orthogonal and there is no entanglement initially. Also, choose an orthonormal basis for H_A, $\{|e_i\rangle\}_i$. (This could be a "continuously indexed basis" or a mixture of continuous and discrete indexes, in which case we would have to use a rigged Hilbert space and be more careful about what we mean by orthonormal but that's an inessential detail for expository purposes.) Then, we can expand

$$U(|\psi_1\rangle \otimes |\mathrm{in}\rangle)$$

And

$$U(|\psi_2\rangle \otimes |\mathrm{in}\rangle)$$

uniquely as

$$\sum_i |e_i\rangle \otimes |f_{1i}\rangle$$

And

$$\sum_i |e_i\rangle \otimes |f_{2i}\rangle$$

respectively. One thing to realize is that the environment contains a huge number of degrees of freedom, a good number of them interacting with each other all the time. This makes the following assumption reasonable in a handwaving way, which can be shown to be true in some simple toy models. Assume that there exists a basis for H_ϵ such that $|f_{1i}\rangle$ and $|f_{1j}\rangle$ are all approximately orthogonal to a good degree if i is not j and the same thing for $|f_{2i}\rangle$ and $|f_{2j}\rangle$ and also $|f_{1i}\rangle$ and $|f_{2j}\rangle$ for any i and j (the decoherence property).

This often turns out to be true (as a reasonable conjecture) in the position basis because how A interacts with the environment would often depend critically upon the position of the objects in A. Then, if we take the partial trace over the environment, we'd find the density state is approximately described by

$$\sum_i (\langle f_{1i}|f_{1i}\rangle + \langle f_{2i}|f_{2i}\rangle)|e_i\rangle\langle e_i|$$

(i.e. we have a diagonal mixed state and there is no constructive or destructive interference and the "probabilities" add up classically). The time it takes for U(t) (the unitary operator as a function of time) to display the decoherence property is called the **decoherence time**.

Experimental Observations

Quantitative Measurement

The decoherence rate depends on a number of factors including temperature, or uncertainty in position, and many experiments have tried to measure it depending on the external environment.

The process of a quantum superposition gradually obliterated by decoherence was quantitatively measured for the first time by Serge Haroche and his co-workers at the École Normale Supérieure in Paris in 1996. Their approach involved sending individual rubidium atoms, each in a superposition of two states, through a microwave-filled cavity. The two quantum states both cause shifts in the phase of the microwave field, but by different amounts, so that the field itself is also put into a superposition of two states. Due to photon scattering on cavity mirror imperfection, the cavity field losses phase coherence to the environment.

Haroche and his colleagues measured the resulting decoherence via correlations between the states of pairs of atoms sent through the cavity with various time delays between the atoms.

Reducing Environmental Decoherence

In July 2011, researchers from University of British Columbia and University of California, Santa Barbara were able to reduce environmental decoherence rate "to levels far below the threshold necessary for quantum information processing" by applying high magnetic fields in their experiment.

In Interpretations of Quantum Mechanics

Before an understanding of decoherence was developed, the Copenhagen interpretation of quantum mechanics treated wavefunction collapse as a fundamental, a priori process. Decoherence provides an explanatory mechanism for the appearance of wavefunction collapse and was first developed by David Bohm in 1952 who applied it to Louis DeBroglie's pilot wave theory, producing Bohmian mechanics, the first successful hidden variables interpretation of quantum mechanics. Decoherence was then used by Hugh Everett in 1957 to form the core of his many-worlds interpretation. However decoherence was largely ignored for many years, and not until the 1980s did decoherent-based explanations of the appearance of wavefunction collapse become popular, with the greater acceptance of the use of reduced density matrices. The range of decoherent interpretations have subsequently been extended around the idea, such as consistent histories. Some versions of the Copenhagen Interpretation have been modified to include decoherence.

Decoherence does not claim to provide a mechanism for the actual wave function collapse; rather it puts forth a reasonable mechanism for the appearance of wavefunction collapse. The quantum nature of the system is simply "leaked" into the environment so that a total superposition of the wavefunction still exists, but exists — at least for all practical purposes — beyond the realm of measurement. Of course by definition the claim that a merged but unmeasurable wavefunction still exists cannot be proven experimentally. Decoherence explains why a quantum system begins to obey classical probability rules after interacting with its environment (due to the suppression of the interference terms when applying Born's probability rules to the system).

QUANTUM ENTANGLEMENT

Quantum entanglement is a physical phenomenon that occurs when pairs or groups of particles are generated or interact in ways such that the quantum state of each particle cannot be described independently of the others, even when the particles are separated by a large distance – instead, a quantum state must be described for the system as a whole.

Measurements of physical properties such as position, momentum, spin, and polarization, performed on entangled particles are found to be appropriately correlated. For example, if a pair of particles are generated in such a way that their total spin is known to be zero, and one particle is found to have clockwise spin on a certain axis, the spin of the other particle, measured on the same axis, will be found to be counterclockwise, as to be expected due to their entanglement. However, this behavior gives rise to paradoxical effects: any measurement of a property of a particle can be seen as acting on that particle (e.g., by collapsing a number of superposed states) and will change the original quantum property by some unknown amount; and in the case of entangled particles, such a measurement will be on the entangled system as a whole. It thus appears that one particle of an entangled pair "knows" what measurement has been performed on the other, and with what outcome, even though there is no known means for such information to be communicated between the particles, which at the time of measurement may be separated by arbitrarily large distances.

Such phenomena were the subject of a 1935 paper by Albert Einstein, Boris Podolsky, and Nathan Rosen,and several papers by Erwin Schrödinger shortly thereafter, describing what came to be known as the EPR paradox. Einstein and others considered such behavior to be impossible, as it violated the local realist view of causality (Einstein referring to it as "spooky action at a distance") and argued that the accepted formulation of quantum mechanics must therefore be incomplete. Later, however, the counterintuitive predictions of quantum mechanics were verified experimentally. Experiments have been performed involving measuring the polarization or spin of entangled particles in different directions, which – by producing violations of Bell's inequality – demonstrate statistically that the local realist view cannot be correct. This has been shown to occur even when the measurements are performed more quickly than light could travel between the sites of measurement: there is no lightspeed or slower influence that can pass between the entangled particles. Recent experiments have measured entangled particles within less than one hundredth of a percent of the travel time of light between them. According to the formalism of quantum theory, the effect of measurement happens instantly. It is not possible, however, to use this effect to transmit classical information at faster-than-light speeds (see Faster-than-light § Quantum mechanics).

Quantum entanglement is an area of extremely active research by the physics community, and its effects have been demonstrated experimentally with photons, neutrinos, electrons, molecules the size of buckyballs, and even small diamonds. Research is also focused on the utilization of entanglement effects in communication and computation.

The counterintuitive predictions of quantum mechanics about strongly correlated systems were first discussed by Albert Einstein in 1935, in a joint paper with Boris Podolsky and Nathan Rosen. In this study, they formulated the EPR paradox (Einstein, Podolsky, Rosen paradox), a thought experiment that attempted to show that quantum mechanical theory was incomplete. They wrote: "We are thus forced to conclude that the quantum-mechanical description of physical reality given by wave functions is not complete."

However, they did not coin the word entanglement, nor did they generalize the special properties of the state they considered. Following the EPR paper, Erwin Schrödinger wrote a letter (in German) to Einstein in which he used the word Verschränkung (translated by himself as entanglement) "to describe the correlations between two particles that interact and then separate, as in the EPR experiment." He shortly thereafter published a seminal paper defining and discussing the notion, and terming it "entanglement." In the paper he recognized the importance of the concept, and stated: "I would not call entanglement one but rather the characteristic trait of quantum mechanics, the one that enforces its entire departure from classical lines of thought."

Like Einstein, Schrödinger was dissatisfied with the concept of entanglement, because it seemed to violate the speed limit on the transmission of information implicit in the theory of relativity. Einstein later famously derided entanglement as "spukhafte Fernwirkung" or "spooky action at a distance."

The EPR paper generated significant interest among physicists and inspired much discussion about the foundations of quantum mechanics (perhaps most famously Bohm's interpretation of quantum mechanics), but produced relatively little other published work. So, despite the interest, the weak point in EPR's argument was not discovered until 1964, when John Stewart Bell proved that one of their key assumptions, the principle of locality, which underlies the kind of hidden variables interpretation hoped for by EPR, was mathematically inconsistent with the predictions of quantum theory. Specifically, he demonstrated an upper limit, seen in Bell's inequality, regarding the strength of correlations that can be produced in any theory obeying local realism, and he showed that quantum theory predicts violations of this limit for certain entangled systems. His inequality is experimentally testable, and there have been numerous relevant experiments, starting with the pioneering work of Stuart Freedman and John Clauser in 1972 and Alain Aspect's experiments in 1982, all of which have shown agreement with quantum mechanics rather than the principle of local realism. Until recently each had left open at least one loophole by which it was possible to question the validity of the results. However, in 2015 the first loophole-free experiment was performed, which ruled out a large class of local realism theories with certainty. The work of Bell raised the possibility of using these super-strong correlations as a resource for communication. It led to the discovery of quantum key distribution protocols, most famously BB84 by Charles H. Bennett and Gilles Brassard and E91 by Artur Ekert. Although BB84 does not use entanglement, Ekert's protocol uses the violation of a Bell's inequality as a proof of security.

Concept

Meaning of Entanglement

An entangled system is defined to be one whose quantum state cannot be factored as a product of states of its local constituents; that is to say, they are not individual particles but are an inseparable whole. In entanglement, one constituent cannot be fully described without considering the other(s). Note that the state of a composite system is always expressible as a sum, or superposition, of products of states of local constituents; it is entangled if this sum necessarily has more than one term.

Quantum systems can become entangled through various types of interactions. For some ways in which entanglement may be achieved for experimental purposes, see the section below on methods. Entanglement is broken when the entangled particles decohere through interaction with the environment; for example, when a measurement is made.

As an example of entanglement: a subatomic particle decays into an entangled pair of other particles. The decay events obey the various conservation laws, and as a result, the measurement outcomes of one daughter particle must be highly correlated with the measurement outcomes of the other daughter particle (so that the total momenta, angular momenta, energy, and so forth remains roughly the same before and after this process). For instance, a spin-zero particle could decay into a pair of spin-½ particles. Since the total spin before and after this decay must be zero (conservation of angular momentum), whenever the first particle is measured to be spin up on some axis, the other, when measured on the same axis, is always found to be spin down. (This is called the spin anti-correlated case; and if the prior probabilities for measuring each spin are equal, the pair is said to be in the singlet state.)

The special property of entanglement can be better observed if we separate the said two particles. Let's put one of them in the White House in Washington and the other in UC Berkeley (think about this as a thought experiment, not an actual one). Now, if we measure a particular characteristic of one of these particles (say, for example, spin), get a result, and then measure the other particle using the same criterion (spin along the same axis), we find that the result of the measurement of the second particle will match (in a complementary sense) the result of the measurement of the first particle, in that they will be opposite in their values.

The above result may or may not be perceived as surprising. A classical system would display the same property, and a hidden variable theory (see below) would certainly be required to do so, based on conservation of angular momentum in classical and quantum mechanics alike. The difference is that a classical system has definite values for all the observables all along while the quantum system does not. In a sense to be discussed below, the quantum system considered here seems to acquire a probability distribution for the outcome of a measurement of the spin along any axis of the other particle upon measurement of the first particle. This probability distribution is in general different from what it would be without measurement of the first particle. This may certainly be perceived as surprising in the case of spatially separated entangled particles.

Paradox

The paradox is that a measurement made on either of the particles apparently collapses the state of the entire entangled system – and does so instantaneously, before any information about the measurement result could have been communicated to the other particle (assuming that information cannot travel faster than light) and hence assured the "proper" outcome of the measurement of the other part of the entangled pair. In the quantum formalism, the result of a spin measurement on one of the particles is a collapse into a state in which each particle has a definite spin (either up or down) along the axis of measurement. The outcome is taken to be random, with each possibility having a probability of 50%. However, if both spins are measured along the same axis, they are found to be anti-correlated. This means that the random outcome of the measurement made on one particle seems to have been transmitted to the other, so that it can make the "right choice" when it too is measured.

The distance and timing of the measurements can be chosen so as to make the interval between the two measurements spacelike, hence, a message connecting the events would have to travel faster than light. According to the principles of special relativity, it is not possible for any information to travel between two such measuring events. It is not even possible to unambiguously say which of the measurements came first. For two spacelike separated events x_1 and x_2 there are inertial systems in which x_1 is first and others in which x_2 is first. Therefore, the correlation between the two measurements cannot appropriately be explained as one measurement determining the other. Different observers would disagree about the role of cause and effect.

In conclusion, the result shows that it is quite impossible for the measurement of one particle's properties to determine the other's properties, even though it seems that way at first, as the speed of light is the universal speed limit, and nothing can travel faster than that speed.

Hidden Variables Theory

A possible resolution to the paradox might be to assume that the state of the particles contains some hidden variables, whose values effectively determine, right from the moment of separation, what the outcomes of the spin measurements are going to be. This would mean that each particle carries all the required information with it, and nothing needs to be transmitted from one particle to the other at the time of measurement. It was originally believed by Einstein and others (see the previous section) that this was the only way out of the paradox, and therefore that the accepted quantum mechanical description (with a random measurement outcome) must be incomplete. (In fact similar paradoxes can arise even without entanglement: the position of a single particle is spread out over space, and two widely separated detectors attempting to detect the particle in two different places must instantaneously attain appropriate correlation, so that they do not both detect the particle.)

Violations of Bell's Inequality

The hidden variables theory fails, however, when we consider measurements of the spin of entangled particles along different axes (for example, along any of three axes which make angles of 120 degrees). If a large number of pairs of such measurements are made (on a large number of pairs of entangled particles), then statistically, if the local realist or hidden variables view were correct, the results would always satisfy Bell's inequality. A number of experiments have shown in practice that Bell's inequality is not satisfied. However, all experiments have loophole problems. When measurements of the entangled particles are made in moving relativistic reference frames, in which each measurement (in its own relativistic time frame) occurs before the other, the measurement results remain correlated.

The fundamental issue about measuring spin along different axes is that these measurements cannot have definite values at the same time—they are incompatible in the sense that these measurements' maximum simultaneous precision is constrained by the uncertainty principle. This is contrary to what is found in classical physics, where any number of properties can be measured simultaneously with arbitrary accuracy. It has been proven mathematically that compatible measurements cannot show Bell-inequality-violating correlations, and thus entanglement is a fundamentally non-classical phenomenon.

Other Types of Experiments

In experiments in 2012 and 2013, polarization correlation was created between photons that never coexisted in time. The authors claimed that this result was achieved by entanglement swapping between two pairs of entangled photons after measuring the polarization of one photon of the early pair, and that it proves that quantum non-locality applies not only to space but also to time.

In three independent experiments in 2013 it was shown that classically-communicated separable quantum states can be used to carry entangled states. The first loophole free Bell test was held in TU Delft in 2015 confirming the violation of Bell inequality.

In August 2014, Brazilian researcher Gabriela Barreto Lemos and team were able to "take pictures" of objects using photons that have not interacted with the subjects, but were entangled with photons that did interact with such objects. Lemos, from the University of Vienna, is confident that this new quantum imaging technique could find application where low light imaging is imperative, in fields like biological or medical imaging.

Mystery of Time

There have been suggestions to look at the concept of time as an emergent phenomenon that is a side effect of quantum entanglement. In other words, time is an entanglement phenomenon, which places all equal clock readings (of correctly prepared clocks – or of any objects usable as clocks) into the same history. This was first fully theorized by Don Page and William Wootters in 1983. The Wheeler–DeWitt equation that combines general relativity and quantum mechanics – by leaving out time altogether – was introduced in the 1960s and it was taken up again in 1983, when the theorists Don Page and William Wootters made a solution based on the quantum phenomenon of entanglement. Page and Wootters argued that entanglement can be used to measure time. In 2013, at the Istituto Nazionale di Ricerca Metrologica (INRIM) in Turin, Italy, researchers performed the first experimental test of Page and Wootters' ideas. Their result has been interpreted to confirm that time is an emergent phenomenon for internal observers but absent for external observers of the universe just as the Wheeler-DeWitt equation predicts.

Source for the Arrow of Time

Physicist Seth Lloyd says that quantum uncertainty gives rise to entanglement, the putative source of the arrow of time. According to Lloyd; "The arrow of time is an arrow of increasing correlations." The approach to entanglement would be from the perspective of the causal arrow of time, with the assumption that the cause of the measurement of one particle determines the effect of the result of the other particle's measurement.

Non-Locality and Entanglement

In the media and popular science, quantum non-locality is often portrayed as being equivalent to entanglement. While it is true that a pure bipartite quantum state must be entangled in order for it to produce non-local correlations, there exist entangled states that do not produce such correlations, and there exist non-entangled (separable) quantum states that present some non-local behaviour. A well-known example of the first case is the Werner state that is entangled for certain values of p_{sym}, but can always be described using local hidden variables. In short, entanglement of a two-party state is necessary but not sufficient for that state to be non-local. Moreover, it was shown that, for arbitrary number of party, there exist states that are genuinely entangled but admits a fully local strategy. It is important to recognize that entanglement is more commonly viewed as an algebraic concept, noted for being a precedent to non-locality as well as to quantum teleportation and to superdense coding, whereas non-locality is defined according to experimental statistics and is much more involved with the foundations and interpretations of quantum mechanics.

Quantum Mechanical Framework

The following subsections are for those with a good working knowledge of the formal, mathematical description of quantum mechanics, including familiarity with the formalism and theoretical framework developed in the articles: bra–ket notation and mathematical formulation of quantum mechanics.

Pure States

Consider two noninteracting systems A and B, with respective Hilbert spaces H_A and H_B. The Hilbert space of the composite system is the tensor product

$$H_A \otimes H_B.$$

If the first system is in state $|\psi\rangle_A$ and the second in state $|\phi\rangle_B$, the state of the composite system is

$$|\psi\rangle_A \otimes |\phi\rangle_B.$$

States of the composite system that can be represented in this form are called separable states, or product states.

Notallstatesareseparablestates(andthusproductstates).Fixabasis $\{|i\rangle_A\}$ for H_A and a basis $\{|j\rangle_B\}$ for H_B. The most general state in $H_A \otimes H_B$ is of the form

$$|\psi\rangle_{AB} = \sum_{i,j} c_{ij} |i\rangle_A \otimes |j\rangle_B.$$

This state is separable if there exist vectors $[c_i^A],[c_j^B]$ so that $c_{ij}=c_i^A c_j^B$, yielding $|\psi\rangle_A=\sum_i c_i^A|i\rangle_A$ and $|\phi\rangle_B=\sum_j c_j^B|j\rangle_B$. It is inseparable if for any vectors $[c_i^A],[c_j^B]$ at least for one pair of coordinates c_i^A,c_j^B we have $c_{ij}\neq c_i^A c_j^B$. If a state is inseparable, it is called an entangled state.

For example, given two basis vectors $\{|0\rangle_A,|1\rangle_A\}$ of H_A and two basis vectors $\{|0\rangle_B,|1\rangle_B\}$ of H_B, the following is an entangled state:

$$\frac{1}{\sqrt{2}}\left(|0\rangle_A \otimes |1\rangle_B - |1\rangle_A \otimes |0\rangle_B\right).$$

If the composite system is in this state, it is impossible to attribute to either system A or system B a definite pure state. Another way to say this is that while the von Neumann entropy of the whole state is zero (as it is for any pure state), the entropy of the subsystems is greater than zero. In this sense, the systems are "entangled". This has specific empirical ramifications for interferometry.[50] It is worthwhile to note that the above example is one of four Bell states, which are (maximally) entangled pure states (pure states of the $H_A \otimes H_B$ space, but which cannot be separated into pure states of each H_A and H_B). Now suppose Alice is an observer for system A, and Bob is an observer for system B. If in the entangled state given above Alice makes a measurement in the $\{|0\rangle,|1\rangle\}$ eigenbasis of A, there are two possible outcomes, occurring with equal probability:[51]

Alice measures 0, and the state of the system collapses to $|0\rangle_A|1\rangle_B$.

Alice measures 1, and the state of the system collapses to $|1\rangle_A|0\rangle_B$

If the former occurs, then any subsequent measurement performed by Bob, in the same basis, will always return 1. If the latter occurs, (Alice measures 1) then Bob's measurement will return 0 with certainty. Thus, system B has been altered by Alice performing a local measurement on system A. This remains true even if the systems A and B are spatially separated. This is the foundation of the EPR paradox.

The outcome of Alice's measurement is random. Alice cannot decide which state to collapse the composite system into, and therefore cannot transmit information to Bob by acting on her system. Causality is thus preserved, in this particular scheme. For the general argument, see no-communication theorem.

Ensembles

As mentioned above, a state of a quantum system is given by a unit vector in a Hilbert space. More generally, if one has a large number of copies of the same system, then the state of this ensemble is described by a density matrix, which is a positive-semidefinite matrix, or a trace class when the state space is infinite-dimensional, and has trace 1. Again, by the spectral theorem, such a matrix takes the general form:

$$\rho = \sum_i w_i |\alpha_i\rangle\langle\alpha_i|,$$

where the w_i are positive-valued probabilities (they sum up to 1), the vectors α_i are unit vectors, and in the infinite-dimensional case, we would take the closure of such states in the trace norm. We can interpret ρ as representing an ensemble where w_i is the proportion of the ensemble whose states are $|\alpha_i\rangle$. When a mixed state has rank 1, it therefore describes a pure ensemble. When there is less than total information about the state of a quantum system we need density matrices to represent the state.

Experimentally, a mixed ensemble might be realized as follows. Consider a "black box" apparatus that spits electrons towards an observer. The electrons' Hilbert spaces are identical. The apparatus might produce electrons that are all in the same state; in this case, the electrons received by the observer are then a pure ensemble. However, the apparatus could produce electrons in different states. For example, it could produce two populations of electrons: one with state $|\mathbf{z}+\rangle$ with spins aligned in the positive **z**direction, and the other with state $|\mathbf{y}-\rangle$ with spins aligned in the negative **y** direction. Generally, this is a mixed ensemble, as there can be any number of populations, each corresponding to a different state.

Following the definition above, for a bipartite composite system, mixed states are just density matrices on $H_A \otimes H_B$. That is, it has the general form

$$\rho = \sum_i w_i \left[\sum_j \bar{c}_{ij}(|\alpha_{ij}\rangle \otimes |\beta_{ij}\rangle)\right] \otimes \left[\sum_k c_{ik}(\langle\alpha_{ik}| \otimes \langle\beta_{ik}|)\right]$$

where the w_i are positively valued probabilities, $\sum_j |c_{ij}|^2 = 1,$ and the vectors are unit vectors. This is self-adjoint and positive and has trace 1.

Extending the definition of separability from the pure case, we say that a mixed state is separable if it can be written as

$$\rho = \sum_i w_i \rho_i^A \otimes \rho_i^B,$$

where the w_i are positively valued probabilities and the ρ_i^A's and ρ_i^B are themselves mixed states (density operators) on the subsystems A and B respectively. In other words, a state is separable if it is a probability distribution over uncorrelated states, or product states. By writing the density matrices as sums of pure ensembles and expanding, we may assume without loss of generality that ρ_i^A's and ρ_i^B are themselves pure ensembles. A state is then said to be entangled if it is not separable.

In general, finding out whether or not a mixed state is entangled is considered difficult. The general bipartite case has been shown to be NP-hard. For the 2 × 2 and 2 × 3 cases, a necessary and sufficient criterion for separability is given by the famous Positive Partial Transpose (PPT) condition.[

Reduced Density Matrices

The idea of a reduced density matrix was introduced by Paul Dirac in 1930. Consider as above systems A and B each with a Hilbert space H_A, H_B. Let the state of the composite system be

$$|\Psi\rangle \in H_A \otimes H_B.$$

As indicated above, in general there is no way to associate a pure state to the component system A. However, it still is possible to associate a density matrix. Let

$$\rho_T = |\Psi\rangle\,\langle\Psi|.$$

which is the projection operator onto this state. The state of A is the partial trace of ρ_T over the basis of system B:

$$\rho_A \stackrel{\text{def}}{=} \sum_j \langle j|_B \left(|\Psi\rangle\langle\Psi|\right)|j\rangle_B = \mathrm{Tr}_B\, \rho_T.$$

ρ_A is sometimes called the reduced density matrix of ρ on subsystem A. Colloquially, we "trace out" system B to obtain the reduced density matrix on A.

For example, the reduced density matrix of A for the entangled state

$$\frac{1}{\sqrt{2}}\left(|0\rangle_A \otimes |1\rangle_B - |1\rangle_A \otimes |0\rangle_B\right),$$

discussed above is

$$\rho_A = \tfrac{1}{2}\left(|0\rangle_A\langle 0|_A + |1\rangle_A\langle 1|_A\right)$$

This demonstrates that, as expected, the reduced density matrix for an entangled pure ensemble is a mixed ensemble. Also not surprisingly, the density matrix of A for the pure product state $|\psi\rangle_A \otimes |\phi\rangle_B$ discussed above is

$$\rho_A = |\psi\rangle_A \langle\psi|_A .$$

In general, a bipartite pure state ρ is entangled if and only if its reduced states are mixed rather than pure.

Two Applications that Use Them

Reduced density matrices were explicitly calculated in different spin chains with unique ground state. An example is the one-dimensional AKLT spin chain: the ground state can be divided into a block and an environment. The reduced density matrix of the block is proportional to a projector to a degenerate ground state of another Hamiltonian.

The reduced density matrix also was evaluated for XY spin chains, where it has full rank. It was proved that in the thermodynamic limit, the spectrum of the reduced density matrix of a large block of spins is an exact geometric sequence in this case.

Entropy

In this section, the entropy of a mixed state is discussed as well as how it can be viewed as a measure of quantum entanglement.

Definition

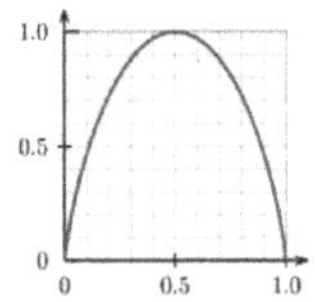

The plot of von Neumann entropy Vs Eigenvalue for a bipartite 2-level pure state. When the eigenvalue has value .5, von Neumann entropy is at a maximum, corresponding to maximum entanglement.

In classical information theory, the Shannon entropy, H is associated to a probability distribution $p_1, \cdots, p_n$ in the following way:

$$H(p_1, \cdots, p_n) = -\sum_i p_i \log_2 p_i .$$

Since a mixed state ρ is a probability distribution over an ensemble, this leads naturally to the definition of the von Neumann entropy:

$$S(\rho) = -\mathrm{Tr}\,(\rho \log_2 \rho) .$$

In general, one uses the Borel functional calculus to calculate a non-polynomial function such as $\log_2(\rho)$. If the nonnegative operator ρ acts on a finite-dimensional Hilbert space and has eigenvalues $\lambda_1, \cdots, \lambda_n$, $\log_2(\rho)$ turns out to be nothing more than the operator with the same eigenvectors, but the eigenvalues $\log_2(\lambda_1), \cdots, \log_2(\lambda_n)$. The Shannon entropy is then:

$$S(\rho) = -\mathrm{Tr}\,(\rho \log_2 \rho) = -\sum_i \lambda_i \log_2 \lambda_i.$$

Since an event of probability 0 should not contribute to the entropy, and given that

$$\lim_{p \to 0} p \log p = 0,$$

the convention 0 log(0) = 0 is adopted. This extends to the infinite-dimensional case as well: if ρ has spectral resolution

$$\rho = \int \lambda dP_\lambda,$$

assume the same convention when calculating

$$\rho \log_2 \rho = \int \lambda \log_2 \lambda dP_\lambda.$$

As in statistical mechanics, the more uncertainty (number of microstates) the system should possess, the larger the entropy. For example, the entropy of any pure state is zero, which is unsurprising since there is no uncertainty about a system in a pure state. The entropy of any of the two subsystems of the entangled state discussed above is log(2) (which can be shown to be the maximum entropy for 2 × 2 mixed states).

As A Measure of Entanglement

Entropy provides one tool which can be used to quantify entanglement, although other entanglement measures exist. If the overall system is pure, the entropy of one subsystem can be used to measure its degree of entanglement with the other subsystems.

For bipartite pure states, the von Neumann entropy of reduced states is the unique measure of entanglement in the sense that it is the only function on the family of states that satisfies certain axioms required of an entanglement measure.

It is a classical result that the Shannon entropy achieves its maximum at, and only at, the uniform probability distribution {1/n,...,1/n}. Therefore, a bipartite pure state ρ ∈ $H_A \otimes H_B$ is said to be a maximally entangled state if the reduced state of ρ is the diagonal matrix

$$\begin{bmatrix} \frac{1}{n} & & \\ & \ddots & \\ & & \frac{1}{n} \end{bmatrix}.$$

For mixed states, the reduced von Neumann entropy is not the only reasonable entanglement measure.

As an aside, the information-theoretic definition is closely related to entropy in the sense of statistical mechanics (comparing the two definitions, we note that, in the present context, it is customary to set the Boltzmann constant k = 1). For example, by properties of the Borel functional calculus, we see that for any unitary operator U,

$$S(\rho) = S\left(U\rho U^*\right).$$

Indeed, without this property, the von Neumann entropy would not be well-defined.

In particular, U could be the time evolution operator of the system, i.e.,

$$U(t) = \exp\left(\frac{-iHt}{\hbar}\right),$$

where H is the Hamiltonian of the system. Here the entropy is unchanged.

The reversibility of a process is associated with the resulting entropy change, i.e., a process is reversible if, and only if, it leaves the entropy of the system invariant. Therefore, the march of the arrow of time towards thermodynamic equilibrium is simply the growing spread of quantum entanglement. This provides a connection between quantum information theory and thermodynamics.

Rényi entropy also can be used as a measure of entanglement.

Entanglement Measures

Entanglement measures quantify the amount of entanglement in a (often viewed as a bipartite) quantum state. As aforementioned, entanglement entropy is the standard measure of entanglement for pure states (but no longer a measure of entanglement for mixed states). For mixed states, there are some entanglement measures in the literature and no single one is standard.

- Entanglement cost
- Distillable entanglement
- Entanglement of formation
- Relative entropy of entanglement
- Squashed entanglement
- Logarithmic negativity

Most (but not all) of these entanglement measures reduce for pure states to entanglement entropy, and are difficult (NP-hard) to compute.

Quantum Field Theory

The Reeh-Schlieder theorem of quantum field theory is sometimes seen as an analogue of quantum entanglement.

Applications

Entanglement has many applications in quantum information theory. With the aid of entanglement, otherwise impossible tasks may be achieved.

Among the best-known applications of entanglement are superdense coding and quantum teleportation.

Most researchers believe that entanglement is necessary to realize quantum computing (although this is disputed by some).

Entanglement is used in some protocols of quantum cryptography. This is because the "shared noise" of entanglement makes for an excellent one-time pad. Moreover, since measurement of either member of an entangled pair destroys the entanglement they share, entanglement-based quantum cryptography allows the sender and receiver to more easily detect the presence of an interceptor.

In interferometry, entanglement is necessary for surpassing the standard quantum limit and achieving the Heisenberg limit.

Entangled States

There are several canonical entangled states that appear often in theory and experiments.

For two qubits, the Bell states are

$$|\Phi^\pm\rangle = \frac{1}{\sqrt{2}}(|0\rangle_A \otimes |0\rangle_B \pm |1\rangle_A \otimes |1\rangle_B)$$

$$|\Psi^\pm\rangle = \frac{1}{\sqrt{2}}(|0\rangle_A \otimes |1\rangle_B \pm |1\rangle_A \otimes |0\rangle_B).$$

These four pure states are all maximally entangled (according to the entropy of entanglement) and form an orthonormal basis (linear algebra) of the Hilbert space of the two qubits. They play a fundamental role in Bell's theorem.

For M>2 qubits, the GHZ state is

$$|\mathrm{GHZ}\rangle = \frac{|0\rangle^{\otimes M} + |1\rangle^{\otimes M}}{\sqrt{2}},$$

which reduces to the Bell state $|\Phi^+\rangle$ for $M = 2$. The traditional GHZ state was defined for $M = 3$.

GHZ states are occasionally extended to qudits, i.e., systems of d rather than 2 dimensions.

Also for M>2 qubits, there are spin squeezed states.[67] Spin squeezed states are a class of squeezed coherent states satisfying certain restrictions on the uncertainty of spin measurements, and are necessarily entangled. Spin squeezed states are good candidates for enhancing precision measurements using quantum entanglement.

For two bosonic modes, a NOON state is

$$|\psi_{\text{NOON}}\rangle = \frac{|N\rangle_a|0\rangle_b + |0\rangle_a|N\rangle_b}{\sqrt{2}},$$

This is like a Bell state $|\Phi^+\rangle$ except the basis kets 0 and 1 have been replaced with "the N photons are in one mode" and "the N photons are in the other mode".

Finally, there also exist twin Fock states for bosonic modes, which can be created by feeding a Fock state into two arms leading to a beam splitter. They are the sum of multiple of NOON states, and can used to achieve the Heisenberg limit.

For the appropriately chosen measure of entanglement, Bell, GHZ, and NOON states are maximally entangled while spin squeezed and twin Fock states are only partially entangled. The partially entangled states are generally easier to prepare experimentally.

Methods of Creating Entanglement

Entanglement is usually created by direct interactions between subatomic particles. These interactions can take numerous forms. One of the most commonly used methods is spontaneous parametric down-conversion to generate a pair of photons entangled in polarisation. Other methods include the use of a fiber coupler to confine and mix photons, the use of quantum dots to trap electrons until decay occurs, the use of the Hong-Ou-Mandel effect, etc., In the earliest tests of Bell's theorem, the entangled particles were generated using atomic cascades.

It is also possible to create entanglement between quantum systems that never directly interacted, through the use of entanglement swapping.

Testing a System for Entanglement

Systems which contain no entanglement are said to be separable. For 2-Qubit and Qubit-Qutrit systems (2×2 and 2×3 respectively) the simple Peres–Horodecki criterion provides both a necessary and a sufficient criterion for separability, and thus for detecting entanglement. However, for the general case, the criterion is merely a sufficient one for separability, as the problem becomes NP-hard. A numerical approach to the problem is suggested by Jon Magne Leinaas, Jan Myrheim and Eirik Ovrum in their paper "Geometrical aspects of entanglement". Leinaas et al. offer a numerical approach, iteratively refining an estimated separable state towards the target state to be tested, and checking if the target state can indeed be reached. An implementation of the algorithm (including a built in Peres-Horodecki criterion testing) is brought in the "StateSeparator" web-app.

China launched the world's first quantum communications satellite: "China has launched the world's first satellite dedicated to testing the fundamentals of quantum communication in space. The \$100m Quantum Experiments at Space Scale (QUESS) mission was launched today Aug 16, 2016 from the Jiuquan Satellite Launch Center in northern China at 01:40 local time. For the next two years, the craft – also named "Micius" after the ancient Chinese philosopher – will demonstrate the feasibility of quantum communication between Earth and space, and test quantum entanglement over unprecedented distances."

Naturally Entangled Systems

The electron shell of multi-electron atoms always consists of entangled electrons. The correct ionization energy can be calculated only by consideration of electron entanglement.

It has been shown by femtosecond transition spectroscopy, that in the photosynthesis of plants, entangled photons exist. An efficient conversion of the photon energy into chemical energy is possible only due to this entanglement.

ENERGY LEVEL

A quantum mechanical system or particle that is bound—that is, confined spatially—can only take on certain discrete values of energy. This contrasts with classical particles, which can have any energy. These discrete values are called energy levels. The term is commonly used for the energy levels of electrons in atoms, ions, or molecules, which are bound by the electric field of the nucleus, but can also refer to energy levels of nuclei or vibrational or rotational energy levels in molecules. The energy spectrum of a system with such discrete energy levels is said to be quantized.

In chemistry and atomic physics, an electron shell, or a principal energy level, may be thought of as an orbit followed by electrons around an atom's nucleus. The closest shell to the nucleus is called the "1 shell" (also called "K shell"), followed by the "2 shell" (or "L shell"), then the "3 shell" (or "M shell"), and so on farther and farther from the nucleus. The shells correspond with the principal quantum numbers (n = 1, 2, 3, 4 ...) or are labeled alphabetically with letters used in the X-ray notation (K, L, M, …).

Each shell can contain only a fixed number of electrons: The first shell can hold up to two electrons, the second shell can hold up to eight (2 + 6) electrons, the third shell can hold up to 18 (2 + 6 + 10) and so on. The general formula is that the nth shell can in principle hold up to $2(n^2)$ electrons. Since electrons are electrically attracted to the nucleus, an atom's electrons will generally occupy outer shells only if the more inner shells have already been completely filled by other electrons. However, this is not a strict requirement: atoms may have two or even three incomplete outer shells. (See Madelung rule for more details.) For an explanation of why electrons exist in these shells see electron configuration. If the potential energy is set to zero at infinite distance from the atomic nucleus or molecule, the usual convention, then bound electron states have negative potential energy. If an atom, ion, or molecule is at the lowest possible energy level, it and its electrons are said to be in the ground state. If it is at a higher energy level, it is said to be excited, or any electrons that have higher energy than the ground state are excited. If more than one quantum mechanical state is at the same energy, the energy levels are "degenerate". They are then called degenerate energy levels.

Explanation

Quantized energy levels result from the relation between a particle's energy and its wavelength. For a confined particle such as an electron in an atom, the wave function has the form of standing waves. Only stationary states with energies corresponding to integral numbers of wavelengths can exist; for other states the waves interfere destructively, resulting in zero probability density. Elementary examples that show mathematically how energy levels come about are the particle in a box and the quantum harmonic oscillator. The electron is a fundamental subatomic particle that carries a negative electric charge.

The first evidence of quantization in atoms was the observation of spectral lines in light from the sun in the early 1800s by Joseph von Fraunhofer and William Hyde Wollaston. The notion of energy levels was proposed in 1913 by Danish physicist Niels Bohr in the Bohr theory of the atom. The modern quantum mechanical theory giving an explanation of these energy levels in terms of the Schrödinger equation was advanced by Erwin Schrödinger and Werner Heisenberg in 1926.

Atoms

Intrinsic Energy Levels

In the formulas for energy of electrons at various levels given below in an atom, the zero point for energy is set when the electron in question has completely left the atom, i.e. when the electron's principal quantum number n = ∞. When the electron is bound to the atom in any closer value of n, the electron's energy is lower and is considered negative.

Orbital State Energy Level: Atom/Ion With Nucleus + One Electron

Assume there is one electron in a given atomic orbital in a hydrogen-like atom (ion). The energy of its state is mainly determined by the electrostatic interaction of the (negative) electron with the (positive) nucleus. The energy levels of an electron around a nucleus are given by :

$$E_n = -hcR_\infty \frac{Z^2}{n^2}$$

(typically between 1 eV and 10^3 eV), where R_∞ is the Rydberg constant, Z is the atomic number, n is the principal quantum number, h is Planck's constant, and c is the speed of light. For hydrogen-like atoms (ions) only, the Rydberg levels depend only on the principal quantum number n.

This equation is obtained from combining the Rydberg formula for any hydrogen-like element (shown below) with $E = h\nu = hc/\lambda$ assuming that the principal quantum number n above = n_1 in the Rydberg formula and $n_2 = \infty$ (principal quantum number of the energy level the electron descends from, when emitting a photon). The Rydberg formula was derived from empirical spectroscopic emission data.

$$\frac{1}{\lambda} = RZ^2\left(\frac{1}{n_1^2} - \frac{1}{n_2^2}\right)$$

An equivalent formula can be derived quantum mechanically from the time-independent Schrödinger equation with a kinetic energy Hamiltonian operator using a wave function as an eigenfunction to obtain the energy levels as eigenvalues, but the Rydberg constant would be replaced by other fundamental physics constants.

Electron-Electron Interactions in Atoms

If there is more than one electron around the atom, electron-electron-interactions raise the energy level. These interactions are often neglected if the spatial overlap of the electron wavefunctions is low.

For multi-electron atoms, interactions between electrons cause the preceding equation to be no longer accurate as stated simply with Z as the atomic number. A simple (though not complete) way to understand this is as a shielding effect, where the outer electrons see an effective nucleus of reduced charge, since the inner electrons are bound tightly to the nucleus and partially cancel its charge. This leads to an approximate correction where Z is substituted with an effective nuclear charge symbolized as Z_{eff} that depends strongly on the principal quantum number.

$$E_{n,l} = -hcR_{\infty}\frac{{Z_{\text{eff}}}^2}{n^2}$$

In such cases, the orbital types (determined by the azimuthal quantum number ℓ) as well as their levels within the molecule affect Z_{eff} and therefore also affect the various atomic electron energy levels. The Aufbau principle of filling an atom with electrons for an electron configuration takes these differing energy levels into account. For filling an atom with electrons in the ground state, the lowest energy levels are filled first and consistent with the Pauli exclusion principle, the Aufbau principle, and Hund's rule.

Fine Structure Splitting

Fine structure arises from relativistic kinetic energy corrections, spin–orbit coupling (an electrodynamic interaction between the electron's spin and motion and the nucleus's electric field) and the Darwin term (contact term interaction of s shell electrons inside the nucleus). These affect the levels by a typical order of magnitude of 10^{-3} eV.

Hyperfine Structure

This even finer structure is due to electron–nucleus spin–spin interaction, resulting in a typical change in the energy levels by a typical order of magnitude of 10^{-4} eV.

Energy Levels Due To External Fields

Zeeman Effect

There is an interaction energy associated with the magnetic dipole moment, $\boldsymbol{\mu}_L$, arising from the electronic orbital angular momentum, L, given by

$$U = -\boldsymbol{\mu}_L \cdot \mathbf{B}$$

with

$$-\boldsymbol{\mu}_L = \frac{e\hbar}{2m}\mathbf{L} = \mu_B \mathbf{L}$$

Additionally taking into account the magnetic momentum arising from the electron spin.

Due to relativistic effects (Dirac equation), there is a magnetic momentum, $\boldsymbol{\mu}_S$, arising from the electron spin

$$-\boldsymbol{\mu}_S = -\mu_B g_S \mathbf{S},$$

with g_S the electron-spin g-factor (about 2), resulting in a total magnetic moment, $\boldsymbol{\mu}$,

$$\boldsymbol{\mu} = \boldsymbol{\mu}_L + \boldsymbol{\mu}_S.$$

The interaction energy therefore becomes

$$U_B = -\boldsymbol{\mu} \cdot \mathbf{B} = \mu_B B(M_L + g_S M_S).$$

Molecules

Chemical bonds between atoms in a molecule form because they make the situation more stable for the involved atoms, which generally means the sum energy level for the involved atoms in the molecule is lower than if the atoms were not so bonded. As separate atoms approach each other to covalently bond, their orbitals affect each other's energy levels to form bonding and antibonding molecular orbitals. The energy level of the bonding orbitals is lower, and the energy level of the antibonding orbitals is higher. For the bond in the molecule to be stable, the covalent bonding electrons occupy the lower energy bonding orbital, which may be signified by such symbols as σ or π depending on the situation. Corresponding anti-bonding orbitals can be signified by adding an asterisk to get σ* or π* orbitals. A non-bonding orbital in a molecule is an orbital with electrons in outer shells which do not participate in bonding and its energy level is the same as that of the constituent atom. Such orbitals can be designated as **n** orbitals. The electrons in an n orbital are typically lone pairs. In polyatomic molecules, different vibrational and rotational energy levels are also involved.

Roughly speaking, a **molecular energy state**, i.e. an eigenstate of the molecular Hamiltonian, is the sum of the electronic, vibrational, rotational, nuclear, and translational components, such that:

$$E = E_{\text{electronic}} + E_{\text{vibrational}} + E_{\text{rotational}} + E_{\text{nuclear}} + E_{\text{translational}}$$

where $E_{electronic}$ is an eigenvalue of the electronic molecular Hamiltonian (the value of the potential energy surface) at the equilibrium geometry of the molecule.

The molecular energy levels are labelled by the molecular term symbols.

The specific energies of these components vary with the specific energy state and the substance.

In molecular physics and quantum chemistry, an **energy level** is a quantized energy of a bound quantum mechanical state.

Energy Level Transitions

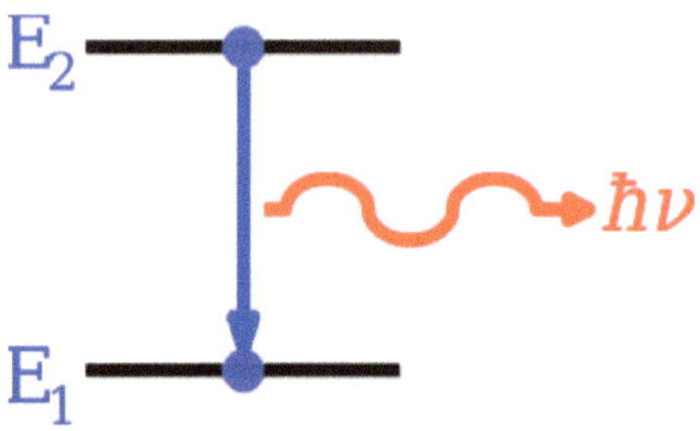

An increase in energy level from E_1 to E_2 resulting from absorption of a photon represented by the red squiggly arrow, and whose energy is $h\nu$

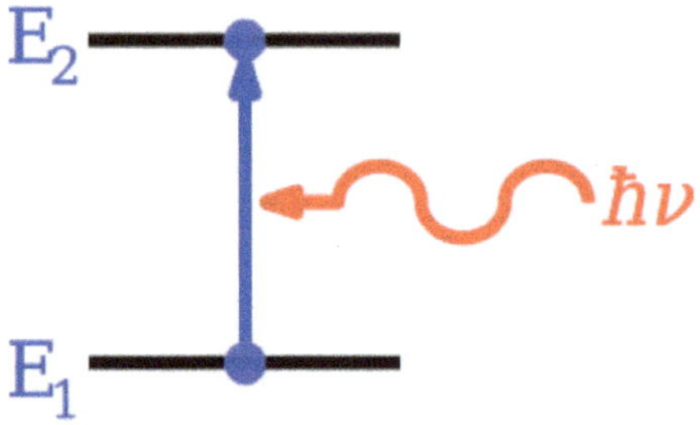

A decrease in energy level from E_2 to E_1 resulting in emission of a photon represented by the red squiggly arrow, and whose energy is $h\nu$

Electrons in atoms and molecules can change (make transitions in) energy levels by emitting or absorbing a photon (of electromagnetic radiation), whose energy must be exactly equal to the energy difference between the two levels. Electrons can also be completely removed from a chemical species such as an atom, molecule, or ion. Complete removal of an electron from an atom can be a form of ionization, which is effectively moving the electron out to an orbital with an infinite principal quantum number, in effect so far away so as to have practically no more effect on the remaining atom (ion). For various types of atoms, there are 1st, 2nd, 3rd, etc. ionization energies for removing the

1st, then the 2nd, then the 3rd, etc. of the highest energy electrons, respectively, from the atom originally in the ground state. Energy in corresponding opposite quantities can also be released, sometimes in the form of photon energy, when electrons are added to positively charged ions or sometimes atoms. Molecules can also undergo transitions in their vibrational or rotational energy levels. Energy level transitions can also be nonradiative, meaning emission or absorption of a photon is not involved.

If an atom, ion, or molecule is at the lowest possible energy level, it and its electrons are said to be in the ground state. If it is at a higher energy level, it is said to be excited, or any electrons that have higher energy than the ground state are excited. Such a species can be excited to a higher energy level by absorbing a photon whose energy is equal to the energy difference between the levels. Conversely, an excited species can go to a lower energy level by spontaneously emitting a photon equal to the energy difference. A photon's energy is equal to Planck's constant (h) times its frequency (f) and thus is proportional to its frequency, or inversely to its wavelength (λ).

$$\Delta E = hf = hc/\lambda,$$

since c, the speed of light, equals to $f\lambda$

Correspondingly, many kinds of spectroscopy are based on detecting the frequency or wavelength of the emitted or absorbed photons to provide information on the material analyzed, including information on the energy levels and electronic structure of materials obtained by analyzing the spectrum.

An asterisk is commonly used to designate an excited state. An electron transition in a molecule's bond from a ground state to an excited state may have a designation such as $\sigma \rightarrow \sigma^*$, $\pi \rightarrow \pi^*$, or $n \rightarrow \pi^*$ meaning excitation of an electron from a σ bonding to a σ antibonding orbital, from a π bonding to a π antibonding orbital, or from an n non-bonding to a π antibonding orbital. Reverse electron transitions for all these types of excited molecules are also possible to return to their ground states, which can be designated as $\sigma^* \rightarrow \sigma$, $\pi^* \rightarrow \pi$, or $\pi^* \rightarrow n$.

A transition in an energy level of an electron in a molecule may be combined with a vibrational transition and called a vibronic transition. A vibrational and rotational transition may be combined by rovibrational coupling. In rovibronic coupling, electron transitions are simultaneously combined with both vibrational and rotational transitions. Photons involved in transitions may have energy of various ranges in the electromagnetic spectrum, such as X-ray, ultraviolet, visible light, infrared, or microwave radiation, depending on the type of transition. In a very general way, energy level differences between electronic states are larger, differences between vibrational levels are intermediate, and differences between rotational levels are smaller, although there can be overlap. Translational energy levels are practically continuous and can be calculated as kinetic energy using classical mechanics.

Higher temperature causes fluid atoms and molecules to move faster increasing their translational energy, and thermally excites molecules to higher average amplitudes of vibrational and rotational modes (excites the molecules to higher internal energy levels). This means that as temperature rises, translational, vibrational, and rotational contributions to molecular heat capacity let molecules absorb heat and hold more internal energy. Conduction of heat typically occurs as molecules or atoms collide transferring the heat between each other. At even higher temperatures, electrons can be thermally excited to higher energy orbitals in atoms or molecules. A subsequent drop of an electron to a lower energy level can release a photon, causing a possibly colored glow.

An electron farther from the nucleus has higher potential energy than an electron closer to the nucleus, thus it becomes less bound to the nucleus, since its potential energy is negative and inversely dependent on its distance from the nucleus.

Crystalline Materials

Crystalline solids are found to have energy bands, instead of or in addition to energy levels. Electrons can take on any energy within an unfilled band. At first this appears to be an exception to the requirement for energy levels. However, as shown in band theory, energy bands are actually made up of many discrete energy levels which are too close together to resolve. Within a band the number of levels is of the order of the number of atoms in the crystal, so although electrons are actually restricted to these energies, they appear to be able to take on a continuum of values. The important energy levels in a crystal are the top of the valence band, the bottom of the conduction band, the Fermi level, the vacuum level, and the energy levels of any defect states in the crystal.

MEASUREMENT IN QUANTUM MECHANICS

The framework of quantum mechanics requires a careful definition of **measurement**. The issue of measurement lies at the heart of the problem of the interpretation of quantum mechanics, for which there is currently no consensus.

Measurement from a Practical Point of View

Measurement plays an important role in quantum mechanics, and it is viewed in different ways among various interpretations of quantum mechanics. In spite of considerable philosophical differences, different views of measurement almost universally agree on the practical question of what results form a routine quantum-physics laboratory measurement. To understand this, the Copenhagen interpretation, which has been commonly used, is employed in this article.

Qualitative Overview

In classical mechanics, a simple system consisting of only one single particle is fully described by the position $\vec{x}(t)$ and momentum $\vec{p}(t)$.

of the particle. As an analogue, in quantum mechanics a system is described by its quantum state, which contains the probabilities of possible positions and momenta. In mathematical language, all possible pure states of a system form an abstract vector space called Hilbert space, which is typically infinite-dimensional. A pure state is represented by a state vector in the Hilbert space. Once a quantum system has been prepared in laboratory, some measurable quantity such as position or energy is measured. For pedagogic reasons, the measurement is usually assumed to be ideally accurate. The state of a system after measurement is assumed to "collapse" into an eigenstate of the operator corresponding to the measurement. Repeating the same measurement without any evolution of the quantum state will lead to the same result. If the preparation is repeated, subsequent measurements will likely lead to different results.

The predicted values of the measurement are described by a probability distribution, or an "average" (or "expectation") of the measurement operator based on the quantum state of the prepared system. The probability distribution is either continuous (such as position and momentum) or discrete (such as spin), depending on the quantity being measured.

The measurement process is often considered as random and indeterministic. Nonetheless, there is considerable dispute over this issue. In some interpretations of quantum mechanics, the result merely appears random and indeterministic, whereas in other interpretations the indeterminism is core and irreducible. A significant element in this disagreement is the issue of "collapse of the wavefunction" associated with the change in state following measurement. There are many philosophical issues and stances (and some mathematical variations) taken—and near universal agreement that we do not yet fully understand quantum reality. In any case, our descriptions of dynamics involve probabilities, not certainties.

Quantitative Details

The mathematical relationship between the quantum state and the probability distribution is, again, widely accepted among physicists, and has been experimentally confirmed countless times. This section summarizes this relationship, which is stated in terms of the mathematical formulation of quantum mechanics.

Measurable Quantities ("Observables") as Operators

It is a postulate of quantum mechanics that all measurements have an associated operator (called an observable operator, or just an observable), with the following properties:

- The observable is a self-adjoint operator mapping a Hilbert space (namely, the state space, which consists of all possible quantum states) into itself.
- Thus, the observable's eigenvectors (called an eigenbasis) form an orthonormal basis that span the state space in which that observable exists. Any quantum state can be represented as a superposition of the eigenstates of an observable.
- Hermitian operators' eigenvalues are real. The possible outcomes of a measurement are precisely the eigenvalues of the given observable.
- For each eigenvalue there are one or more corresponding eigenvectors (eigenstates). A measurement results in the system being in the eigenstate corresponding to the eigenvalue result of the measurement. If the eigenvalue determined from the measurement corresponds to more than one eigenstate ("degeneracy"), instead of being in a definite state, the system is in a sub-space of the measurement operator corresponding to all the states having that eigenvalue.

Important examples of observables are:

The Hamiltonian operator $\hat{H}$, which represents the total energy of the system. In nonrelativistic quantum mechanics the nonrelativistic Hamiltonian operator is given by

$$\hat{H} = \hat{T} + \hat{V} = \frac{\hat{p}^2}{2m} + V(\hat{x}).$$

The momentum operator $\hat{p}$ is given by $\hat{p} = -i\hbar\frac{\partial}{\partial x}$ (in the position basis),or $\hat{p} = p$ (in the momentum basis).

The position operator $\hat{x}$ is given by $\hat{x} = x$ (in the position basis), or $\hat{x} = i\hbar\frac{\partial}{\partial p}$ (in the momentum basis).

Operators can be noncommuting. Two Hermitian operators commute if (and only if) there is at least one basis of vectors such that each of which is an eigenvector of both operators (this is sometimes called a **simultaneous eigenbasis**). Noncommuting observables are said to be incompatible and cannot in general be measured simultaneously. In fact, they are related by an uncertainty principle as discovered by Werner Heisenberg.

Measurement Probabilities and Wavefunction Collapse

There are a few possible ways to mathematically describe the measurement process (both the probability distribution and the collapsed wavefunction). The most convenient description depends on the spectrum (i.e., set of eigenvalues) of the observable.

Discrete, Nondegenerate Spectrum

Let $\hat{O}$ be an observable. By assumption, $\hat{O}$ has discrete eigenstates $|1\rangle, |2\rangle, |3\rangle, \ldots$ with corresponding distinct eigenvalues $O_1, O_2, O_3, \ldots$. hat is, the states are nondegenerate.

Consider a system prepared in state $|\psi\rangle$. Since the eigenstates of the observable $\hat{O}$ form a complete basis called eigenbasis, the state vector $|\psi\rangle$. can be written in terms of the eigenstates as

$$|\psi\rangle = c_1|1\rangle + c_2|2\rangle + c_3|3\rangle + \cdots = \sum_n c_n|n\rangle.$$

Where $c_1, c_2, \ldots$ are complex numbers in general. The eigenvalues $O_1, O_2, O_3, \ldots$ are all possible values of the measurement. The corresponding probabilities are given by

$$\Pr(O_n) = \frac{|\langle n|\psi\rangle|^2}{|\langle\psi|\psi\rangle|} = \frac{|c_n|^2}{\sum_k |c_k|^2}$$

Usually $|\psi\rangle$ is assumed to be normalized, i.e. $\langle\psi|\psi\rangle = 1$. Therefore, the expression above is reduced to

$$\Pr(O_n) = |\langle n|\psi\rangle|^2 = |c_n|^2$$

If the result of the measurement is o_n, then the system (after measurement) is in pure state $|n\rangle$. That is,

$$|\psi'\rangle = |n\rangle$$

so any repeated measurement of $\hat{O}$ will yield the same result o_n. When there is a discontinuous change in state due to a measurement that involves discrete eigenvalues, that is called wave function collapse. For some, this is simply a description of a reasonably accurate discontinuous change in a mathematical representation of physical reality; for others, depending on philosophical orientation, this is a fundamentally serious problem with quantum theory; others see this as statistically-justified approximation resulting from the fact that the entity performing this measurement has been excluded from the state-representation. In particular, multiple measurements of certain physically extended systems demonstrate predicted statistical correlations which would not be possible under classical assumptions.

Continuous, Nondegenerate Spectrum

Let $\hat{O}$ be an observable. By assumption, $\hat{O}$ has continuous eigenstate $|x\rangle$, with corresponding distinct eigenvalue x. The eigenvalue forms a continuous spectrum filling the interval (a,b).

Consider a system prepared in state $|\psi\rangle$. Since the eigenstates of the observable $\hat{O}$ form a complete basis called eigenbasis, the state vector $|\psi\rangle$. can be written in terms of the eigenstates as

$$|\psi\rangle = \int_a^b c(x)|x\rangle\, dx,$$

Where $c(x)$ is a complex-valued function. The eigenvalue that fills up the interval (a,b) is the possible value of measurement. The corresponding probability is described by a probability function given by

$$\Pr(d < x < e) = \frac{\int_d^e |\langle x|\psi\rangle|^2\, dx}{\int_a^b \langle\psi|\psi\rangle\, dx} = \frac{\int_d^e |c(x)|^2\, dx}{\int_a^b |c(x)|^2\, dx}$$

Where $(d,e) \subseteq (a,b)$. Usually $|\psi\rangle$ is assumed to be normalized, i.e. $\int_a^b \langle\psi|\psi\rangle\, dx = 1$. Therefore, the expression above is reduced to

$$\Pr(d < x < e) = \int_d^e |c(x)|^2\, dx$$

If the result of the measurement is x, then the system (after measurement) is in pure state, $x\rangle$ That is,

$$|\psi'\rangle = |x\rangle.$$

Alternatively, it is often possible and convenient to analyze a continuous-spectrum measurement by taking it to be the limit of a different measurement with a discrete spectrum. For example, an analysis of scattering involves a continuous spectrum of energies, but by adding a "box" potential (which bounds the volume in which the particle can be found), the spectrum becomes discrete. By considering larger and larger boxes, this approach need not involve any approximation, but rather can be regarded as an equally valid formalism in which this problem can be analyzed.

Degenerate Spectra

If there are multiple eigenstates with the same eigenvalue (called degeneracies), the analysis is a bit less simple to state, but not essentially different. In the discrete case, for example, instead of finding a complete eigenbasis, it is a bit more convenient to write the Hilbert space as a direct sum of multiple eigenspaces. The probability of measuring a particular eigenvalue is the squared component of the state vector in the corresponding eigenspace, and the new state after measurement is the projection of the original state vector into the appropriate eigenspace.

Density Matrix Formulation

Instead of performing quantum-mechanics computations in terms of wavefunctions (kets), it is sometimes necessary to describe a quantum-mechanical system in terms of a density matrix. The analysis in this case is formally slightly different, but the physical content is the same, and indeed this case can be derived from the wavefunction formulation above. The result for the discrete, degenerate case, for example, is as follows:

Let $\hat{o}$ be an observable, and suppose that it has discrete eigenvalues $O_1, O_2, O_3, \ldots$, associated with eigenspaces $V_1, V_2, \ldots$ respectively. Let P_n be the projection operator into the space V_n.

Assume the system is prepared in the state described by the density matrix ρ. Then measuring $\hat{O}$ can yield any of the results $O_1, O_2, O_3, \ldots$, with corresponding probabilities given by

$$\Pr(O_n) = \mathrm{Tr}(P_n \rho)$$

where T_r denotes trace. If the result of the measurement is n, then the new density matrix will be

$$\rho' = \frac{P_n \rho P_n}{\mathrm{Tr}(P_n \rho)}$$

Alternatively, one can say that the measurement process results in the new density matrix

$$\rho'' = \sum_n P_n \rho P_n$$

where the difference is that ρ'' is the density matrix describing the entire ensemble, whereas ρ' is the density matrix describing the sub-ensemble whose measurement result was.n

Statistics of Measurement

As detailed above, the result of measuring a quantum-mechanical system is described by a probability distribution. Some properties of this distribution are as follows:

Suppose we take a measurement corresponding to observable $\hat{O}$, on a state whose quantum state is $|\psi\rangle$.

- The mean (average) value of the measurement is

$\langle\psi|\hat{O}|\psi\rangle$

- The variance of the measurement is

$\langle\psi|\hat{O}^2|\psi\rangle - (\langle\psi|\hat{O}|\psi\rangle)^2$

- The standard deviation of the measurement is

$\sqrt{\langle\psi|\hat{O}^2|\psi\rangle - (\langle\psi|\hat{O}|\psi\rangle)^2}$

These are direct consequences of the above formulas for measurement probabilities.

Wavefunction Collapse

The process in which a quantum state becomes one of the eigenstates of the operator corresponding to the measured observable is called "collapse", or "wavefunction collapse". The final eigenstate appears randomly with a probability equal to the square of its overlap with the original state. The process of collapse has been studied in many experiments, most famously in the double-slit experiment. The wavefunction collapse raises serious questions regarding "the measurement problem", as well as questions of determinism and locality, as demonstrated in the EPR paradox and later in GHZ entanglement. (See below.)

In the last few decades, major advances have been made toward a theoretical understanding of the collapse process. This new theoretical framework, called quantum decoherence, supersedes previous notions of instantaneous collapse and provides an explanation for the absence of quantum coherence after measurement. Decoherence correctly predicts the form and probability distribution of the final eigenstates, and explains the apparent randomness of the choice of final state in terms of einselection.

Von Neumann Measurement Scheme

The von Neumann measurement scheme, the ancestor of quantum decoherence theory, describes measurements by taking into account the measuring apparatus which is also treated as a quantum object.

"Measurement" of the First Kind — Premeasurement Without Detection

Let the quantum state be in the superposition $|\psi\rangle=\sum_n c_n|\psi_n\rangle$, where $|\psi_n\rangle$ are eigenstates of the operator for the so-called «measurement» prior to von Neumann›s second apparatus. In order to make the «measurement», the system described by $|\psi\rangle$ needs to interact with the measuring apparatus described by the quantum state $|\phi\rangle$, so that the total wave function before the measurement and interaction with the second apparatus is $|\psi\rangle|\phi\rangle$. During the interaction of object and measuring instrument the unitary evolution is supposed to realize the following transition from the initial to the final total wave function:

$$|\psi\rangle|\phi\rangle \to \sum_n c_n|\psi_n\rangle|\phi_n\rangle \quad \textbf{(measurement of the first kind)},$$

Where $|\phi_n\rangle$ are orthonormal states of the measuring apparatus. The unitary evolution above is referred to as **premeasurement**. The relation with wave function collapse is established by calculating the final density operator of the object $\sum_n |c_n|^2|\psi_n\rangle\langle\psi_n|$ from the final total wave function. This density operator is interpreted by von Neumann as describing an ensemble of objects being after the measurement with probability $|c_n|^2$ in the state $|\psi_n\rangle$. The transition

$$|\psi\rangle \to \sum_n |c_n|^2|\psi_n\rangle\langle\psi_n|\psi\rangle$$

is often referred to as weak von Neumann projection, the wave function collapse or strong von Neumann projection

$$|\psi\rangle \to \sum_n |c_n|^2|\psi_n\rangle\langle\psi_n|\psi\rangle \to |\psi_n\rangle$$

being thought to correspond to an additional selection of a subensemble by means of observation.

In case the measured observable has a degenerate spectrum, weak von Neumann projection is generalized to Lüders projection

$$|\psi\rangle \to \sum_n |c_n|^2 P_n, \; P_n = \sum_i |\psi_{ni}\rangle\langle\psi_{ni}|,$$

in which the vectors $|\psi_{ni}\rangle$ for fixed n are the degenerate eigenvectors of the measured observable. For an arbitrary state described by a density operator p Lüders projection is given by

$$\rho \to \sum_n P_n \rho P_n.$$

Measurement of the Second Kind — With Irreversible Detection

In a measurement of the second kind the unitary evolution during the interaction of object and measuring instrument is supposed to be given by

$$|\psi\rangle|\phi\rangle \to \sum_n c_n|\chi_n\rangle|\phi_n\rangle,$$

in which the states $|\chi_n\rangle$ of the object are determined by specific properties of the interaction between object and measuring instrument. They are normalized but not necessarily mutually orthogonal. The relation with wave function collapse is analogous to that obtained for measurements of the first kind, the final state of the object now being $|\chi_n\rangle$ with probability $|c_n|^2$. Note that many measurement procedures are measurements of the second kind, some even functioning correctly only as a consequence of being of the second kind. For instance, a photon counter, detecting a photon by absorbing and hence annihilating it, thus ideally leaving the electromagnetic field in the vacuum state rather than in the state corresponding to the number of detected photons; also the Stern–Gerlach experiment would not function at all if it really were a measurement of the first kind.

Decoherence in Quantum Measurement

One can also introduce the interaction with the environment $|e\rangle$, so that, in a measurement of the first kind, after the interaction the total wave function takes a form

$$\sum_n c_n |\psi_n\rangle |\phi_n\rangle |e_n\rangle,$$

which is related to the phenomenon of decoherence.

The above is completely described by the Schrödinger equation and there are not any interpretational problems with this. Now the problematic wavefunction collapse does not need to be understood as a process $|\psi\rangle \rightarrow |\psi_n\rangle$ on the level of the measured system, but can also be understood as a process $|\psi\rangle \rightarrow |\psi_n\rangle$ on the level of the measured system, but can also be understood as a process $|\phi\rangle \rightarrow |\phi_n\rangle$ on the level of the environment. Studying these processes provides considerable insight into the measurement problem by avoiding the arbitrary boundary between the quantum and classical worlds, though it does not explain the presence of randomness in the choice of final eigenstate. If the set of states

$$\{|\psi_n\rangle\}, \{|\phi_n\rangle\}, \text{ or } \{|e_n\rangle\}$$

represents a set of states that do not overlap in space, the appearance of collapse can be generated by either the Bohm interpretation or the Everett interpretation which both deny the reality of wavefunction collapse. Both of these are stated to predict the same probabilities for collapses to various states as the conventional interpretation by their supporters. The Bohm interpretation is held to be correct only by a small minority of physicists, since there are difficulties with the generalization for use with relativistic quantum field theory. However, there is no proof that the Bohm interpretation is inconsistent with quantum field theory, and work to reconcile the two is ongoing. The Everett interpretation easily accommodates relativistic quantum field theory.

Interaction without Interaction in Quantum Measurement

Interaction without interaction is a new quantum mechanical measurement effect, which states that one motion A does not interact with the other motion B in a system, when we measure the physical quantity associated with the motion A, the other motion B will have an effect on the measured physical quantity. The coinage follows from John

Wheeler's style, for example Wheeler's coinages include 'mass without mass', 'charge without charge' and 'law without law'.

The motion B is usually a classical harmonic vibration, two measured quantities associated with the motions A are proposed i.e. the quantum entanglement of the two two-level atoms in a single-mode polarized cavity field and the number of atoms in an atomic beam reaching the atomic detector. Because of the non-unity trace over the state of a classical harmonic oscillator, during a shorter time interval Δt than its period T, the measured entanglement concurrence between the two atoms is modified by the vibrant factor $(\Delta t)^2/T^2$ the registered number of atoms of the translational motion should be multiplied by another vibrant factor $\Delta t/T$

Actually if the Hamiltonian of the system is given by $H=H_A$ in which the coupling term between the motion A and the motion B is absent then the state of the system read $\rho = \rho_A \otimes \rho_B$ The measure quantity belongigg to the motion A should be (a)= $Tr_a(p_aI_a)$ • $Tr_B(p_BI_B)$ usually the trace over the motion B $Tr_B(p_BI_B)$ is unity our conventional intuition hold. However in some condition the trace over the motion B $Tr_B(p_BI_B)$ for instance a classical harmonic vibrating during a shorter time interval than its period, Should less then unity, then the measurement effect of interaction without interaction appears.

Surprisingly and interestingly the measurement effect for an atomic beam is also a macroscopic quantum phenomenon, because the classical harmonic vibration and the process of registering the number of atoms by the atomic detector are both regarded as the macroscopic events. The measurement effect for an atomic beam is potentially important in the detection of gravitational waves, because the vibrant factor $\Delta t/T$ is independent of the amplitude and the initial phase, which implies that an atomic beam can be used to detect the extremely weak classical harmonic vibrations induced by gravitational waves.

Philosophical Problems of Quantum Measurements

What Physical Interaction Constitutes a Measurement?

Until the advent of quantum decoherence theory in the late 20th century, a major conceptual problem of quantum mechanics and especially the Copenhagen interpretation was the lack of a distinctive criterion for a given physical interaction to qualify as "a measurement" and cause a wavefunction to collapse. This is illustrated by the Schrödinger's cat paradox. Certain aspects of this question are now well understood in the framework of quantum decoherence theory, such as an understanding of weak measurements, and quantifying what measurements or interactions are sufficient to destroy quantum coherence. Nevertheless, there remains less than universal agreement among physicists on some aspects of the question of what constitutes a measurement.

Does Measurement Actually Determine the State?

The question of whether (and in what sense) a measurement actually determines the state is one which differs among the different interpretations of quantum mechanics. (It is also closely related to the understanding of wavefunction collapse.) For example, in most versions of the Copenhagen interpretation, the measurement determines the state, and after measurement the state is definitely what was measured. But according to the many-worlds interpretation, measurement determines the state in a more restricted sense: In other "worlds", other measurement results were obtained, and the other possible states still exist.

Is the Measurement Process Random or Deterministic?

As described above, there is universal agreement that quantum mechanics appears random, in the sense that all experimental results yet uncovered can be predicted and understood in the framework of quantum mechanics measurements being fundamentally random. Nevertheless, it is not settled whether this is true, fundamental randomness, or merely "emergent" randomness resulting from underlying hidden variables which deterministically cause measurement results to happen a certain way each time. This continues to be an area of active research.

If there are hidden variables, Bell's theorem suggests they would have to be "nonlocal".

Does the Measurement Process Violate Locality?

In physics, the **Principle of locality** is the concept that information cannot travel faster than the speed of light (also see special relativity). It is known experimentally (see Bell's theorem, which is related to the EPR paradox) that if quantum mechanics is deterministic (due to hidden variables, as described above), then it is **nonlocal** (i.e. violates the principle of locality). Nevertheless, there is not universal agreement among physicists on whether quantum mechanics is nondeterministic, nonlocal, or both.

REFERENCES

1. Hall, George M. (1997). The Ingenious Mind of Nature: Deciphering the Patterns of Man, Society, and the. Springer. p. 409. ISBN 978-0-306-45571-1.
2. Whitaker, Andrew (2006). Einstein, Bohr and the Quantum Dilemma: From Quantum Theory to Quantum Dillema. Cambridge. p. 414. ISBN 9780521671026.
3. Selleri, Franco (2012). Wave-Particle Duality. Springer. p. 55. ISBN 978-1461364689.
4. Walker, Evan Harris (2000). The Physics of Consciousness. Cambridge, Massachusetts: Perseus. p. 271. ISBN 0-7382-0436-6. ...the founders of quantum mechanics – Heisenberg, Schrödinger and Bohr...
5. Cramer, J. G. et al. "An Inquiry into the Possibility of Nonlocal Quantum Communication". Foundation of Phys. Submitted. Revised Feb., 2015
6. http://plato.stanford.edu/entries/bohr-correspondence/
7. Jim Baggott (2011). The Quantum Story: A History in. Oxford University Press. p. 97.
8. Boscá Díaz-Pintado, María C. (29–31 March 2007). "Updating the wave-particle duality". 15th UK and European Meeting on the Foundations of Physics. Leeds,

UK. Retrieved 2008-06-21.

9. F. A. M. Frescura, B. J. Hiley: Algebras, quantum theory and pre-space, published in Revista Brasileira de Fisica, Volume Especial, Julho 1984, Os 70 anos de Mario Schonberg, pp. 49–86, p. 2
10. Jørgen Kalckar, Niels Bohr, Léon Rosenfeld, Erik Rüdinger, Finn Aaserud (1996). Foundations of Quantum Physics II (1933-1958). Elsevier. p. 210. ISBN 978-0-444-89892-0. Retrieved 2011-10-24.
11. Bandyopadhyay, Supriyo (2000). "Welcher Weg Experiments and the Orthodox Bohr's Complementarity Principle". Physics Letters A. 276 (5–6): 233–239. arXiv:quant-ph/0003073 . Bibcode:2000PhLA..276..233B. doi:10.1016/S0375-9601(00)00670-8.
12. Niels Bohr; fwd. Léon Rosenfeld; ed. Kalckar; et al. (1996). "Complementarity: Bedrock of the Quantal Description". Foundations of Quantum Physics II (1933–1958). Niels Bohr Collected Works. 7. Elsevier. pp. 284–285. ISBN 978-0-444-89892-0.
13. John Wheeler, Physics Today, January 1963, p. 30.
14. Haroche, Serge; Raimond, Jean-Michel (2006). Exploring the Quantum: Atoms, Cavities, and Photons (1st ed.). Oxford University Press. ISBN 978-0198509141.
15. Colella, R.; Overhauser, A. W.; Werner, S. A. (1975). "Observation of gravitationally induced quantum interference" (PDF). Phys. Rev. Lett. 34 (23): 1472–1474. Bibcode:1975PhRvL..34.1472C. doi:10.1103/physrevlett.34.1472.
16. Helmut Rauch; Samuel A. Werner (2000). Neutron Interferometry: Lessons in Experimental Quantum Mechanics. Oxford University Press. ISBN 978-0-19-850027-8.
17. Bohr N (1928). "The Quantum Postulate and the Recent Development of Atomic Theory". Nature. 121: 580–590. Bibcode:1928Natur.121..580B. doi:10.1038/121580a0. Available in the collection of Bohr's early writings, Atomic Theory and the Description of Nature (1934).
18. Niels Bohr (1949). "Discussions with Einstein on Epistemological Problems in Atomic Physics". In P. Schilpp. Albert Einstein: Philosopher-Scientist. Open Court.
19. Saunders S (2005). "Complementarity and Scientific Rationality". Foundations of Physics. 35 (3): 417–447. arXiv:quant-ph/0412195 . Bibcode:2005FoPh...35..417S. doi:10.1007/s10701-004-1982-x.
20. Schlosshauer, Maximilian (2005). "Decoherence, the measurement problem, and interpretations of quantum mechanics". Reviews of Modern Physics. 76 (4): 1267–1305. arXiv:quant-ph/0312059 . Bibcode:2004RvMP...76.1267S. doi:10.1103/RevModPhys.76.1267.
21. Bacon, D. (2001). "Decoherence, control, and symmetry in quantum computers". arXiv:quant-ph/0305025 .
22. Lidar, Daniel A.; Whaley, K. Birgitta (2003). "Decoherence-Free Subspaces and Subsystems". In Benatti, F.; Floreanini, R. Irreversible Quantum Dynamics. Springer Lecture Notes in Physics. 622. Berlin. pp. 83–120. arXiv:quant-

ph/0301032 .

23. Wojciech H. Zurek, Decoherence, einselection, and the quantum origins of the classical, Reviews of Modern Physics 2003, 75, 715 or http://arxiv.org/abs/quant-ph/0105127
24. Wojciech H. Zurek, Decoherence and the transition from quantum to classical, Physics Today, 44, pp 36–44 (1991)
25. Zurek, Wojciech (2002). "Decoherence and the Transition from Quantum to Classical—Revisited" (PDF). Los Alamos Science. 27.
26. A. S. Sanz, F. Borondo: A quantum trajectory description of decoherence, quant-ph/0310096v5
27. V. P. Belavkin (1989). "A new wave equation for a continuous non-demolition measurement". Physics Letters A. 140 (7-8): 355–358. arXiv:quant-ph/0512136 . Bibcode:1989PhLA..140..355B. doi:10.1016/0375-9601(89)90066-2.
28. Howard J. Carmichael (1993). An Open Systems Approach to Quantum Optics. Berlin Heidelberg New-York: Springer-Verlag.
29. Michel Bauer; Denis Bernard; Tristan Benoist. Iterated Stochastic Measurements (Technical report). arXiv:1210.0425 .
30. Decoherence-free subspaces for quantum computation from arXiv
31. Dan Stahlke. "Quantum Decoherence and the Measurement Problem" (PDF). Retrieved 2011-07-23.
32. Serge Haroche; et al. (9 December 1996). "Observing the Progressive Decoherence of the "Meter" in a Quantum Measurement". Phys. Rev. Lett. 77 (24): 4887. Bibcode:1996PhRvL..77.4887B. doi:10.1103/PhysRevLett.77.4887.
33. "Discovery may overcome obstacle for quantum computing: UBC, California researchers". University of British Columbia. 2011-07-20. Retrieved 2011-07-23. Our theory also predicted that we could suppress the decoherence, and push the decoherence rate in the experiment to levels far below the threshold necessary for quantum information processing, by applying high magnetic fields. (...) Magnetic molecules now suddenly appear to have serious potential as candidates for quantum computing hardware," said Susumu Takahashi, assistant professor of chemistry and physics at the University of Southern California. "This opens up a whole new area of experimental investigation with sizeable potential in applications, as well as for fundamental work."
34. "USC Scientists Contribute to a Breakthrough in Quantum Computing". University of California, Santa Barbara. 2011-07-20. Retrieved 2011-07-23.
35. "Breakthrough removes major hurdle for quantum computing". ZDNet. 2011-07-20. Retrieved 2011-07-23.
36. David Bohm, A Suggested Interpretation of the Quantum Theory in Terms of "Hidden Variables", I, Physical Review, (1952), 85, pp 166–179
37. David Bohm, A Suggested Interpretation of the Quantum Theory in Terms of "Hidden Variables", II, Physical Review, (1952), 85, pp 180–193
38. Hugh Everett, Relative State Formulation of Quantum Mechanics, Reviews of

Modern Physics vol 29, (1957) pp 454–462.

39. H. Dieter Zeh, On the Interpretation of Measurement in Quantum Theory, Foundation of Physics, vol. 1, pp. 69-76, (1970).
40. Wojciech H. Zurek, Pointer Basis of Quantum Apparatus: Into what Mixture does the Wave Packet Collapse?, Physical Review D, 24, pp. 1516–1525 (1981)
41. Wojciech H. Zurek, Environment-Induced Superselection Rules, Physical Review D, 26, pp.1862–1880, (1982)
42. Roger Penrose The Road to Reality, p 802-803. "...the environmental-decoherence viewpoint..maintains that state vector reduction the R process can be understood as coming about because the environmental system under consideration becomes inextricably entangled with its environment.... We think of the environment as extremely complicated and essentially 'random' .., accordingly we sum over the unknown states in the environment to obtain a density matrix... Under normal circumstances, one must regard the density matrix as some kind of approximation to the whole quantum truth. For there is no general principle providing an absolute bar to extracting information from the environment.... Accordingly, such descriptions are referred to as FAPP For All Practical Purposes"
43. Huw Price, Times' Arrow and Archimedes' Point p 226. 'There is a world of difference between saying "the environment explains why collapse happens where it does" and saying "the environment explains why collapse seems to happen even though it doesn't really happen'.
44. Einstein A, Podolsky B, Rosen N; Podolsky; Rosen (1935). "Can Quantum-Mechanical Description of Physical Reality Be Considered Complete?". Phys. Rev. 47 (10): 777–780. Bibcode:1935PhRv...47..777E. doi:10.1103/PhysRev.47.777.
45. Schrödinger E (1935). "Discussion of probability relations between separated systems". Mathematical Proceedings of the Cambridge Philosophical Society. 31 (4): 555–563. Bibcode:1935PCPS...31..555S. doi:10.1017/S0305004100013554.
46. Schrödinger E (1936). "Probability relations between separated systems". Mathematical Proceedings of the Cambridge Philosophical Society. 32 (3): 446–452. Bibcode:1936PCPS...32..446S. doi:10.1017/S0305004100019137.
47. Physicist John Bell depicts the Einstein camp in this debate in his article entitled "Bertlmann's socks and the nature of reality", p. 143 of Speakable and unspeakable in quantum mechanics: "For EPR that would be an unthinkable 'spooky action at a distance'. To avoid such action at a distance they have to attribute, to the space-time regions in question, real properties in advance of observation, correlated properties, which predetermine the outcomes of these particular observations. Since these real properties, fixed in advance of observation, are not contained in quantum formalism, that formalism for EPR is incomplete. It may be correct, as far as it goes, but the usual quantum formalism cannot be the whole story." And again on p. 144 Bell says: "Einstein had no difficulty accepting that affairs in different places could be correlated. What he could not accept was that an intervention at one place could influence, immediately, affairs at the other." Downloaded 5 July 2011 from Bell, J. S. (1987). Speakable

and Unspeakable in Quantum Mechanics (PDF). CERN. ISBN 0521334950. Retrieved 2014-06-14.

48. "75 years of entanglement – Science News". Retrieved 13 October 2014.
49. Francis, Matthew. Quantum entanglement shows that reality can't be local, Ars Technica, 30 October 2012
50. Juan Yin; et al. (2013). "Bounding the speed of 'spooky action at a distance". Phys. Rev. Lett. 110, 260407. 1303: 614. arXiv:1303.0614 . Bibcode:2013arXiv1303.0614Y.
51. Matson, John (13 August 2012). "Quantum teleportation achieved over record distances". Nature.
52. Griffiths, David J. (2004), Introduction to Quantum Mechanics (2nd ed.), Prentice Hall, ISBN 0-13-111892-7
53. Roger Penrose, The Road to Reality: A Complete Guide to the Laws of the Universe, London, 2004, p. 603.
54. "New High-Intensity Source of Polarization-Entangled Photon Pairs". Physical Review Letters. 75: 4337–4341. Bibcode:1995PhRvL..75.4337K. doi:10.1103/PhysRevLett.75.4337.
55. "Experimental demonstration of five-photon entanglement and open-destination teleportation". Nature. 430: 54–58. July 2004. arXiv:quant-ph/0402096 . Bibcode:2004Natur.430...54Z. doi:10.1038/nature02643. PMID 15229594.
56. "Experimental entanglement of six photons in graph states". Nature Physics. 3: 91–95. arXiv:quant-ph/0609130 . Bibcode:2007NatPh...3...91L. doi:10.1038/nphys507.
57. "Observation of eight-photon entanglement". Nature Photonics. 6: 225–228. arXiv:1105.6318 . Bibcode:2012NaPho...6..225Y. doi:10.1038/nphoton.2011.354.
58. J. A. Formaggio, D. I. Kaiser, M. M. Murskyj, and T. E. Weiss (2016), "Violation of the Leggett-Garg inequality in neutrino oscillations". Phys. Rev. Lett. Accepted 23 June 2016. https://journals.aps.org/prl/accepted/6f072Y00C3318d41f5739ec7f83a9acf1ad67b002
59. Hensen, B.; et al. (21 October 2015). "Loophole-free Bell inequality violation using electron spins separated by 1.3 kilometres". Nature. 526: 682–686. Bibcode:2015Natur.526..682H. doi:10.1038/nature15759. Retrieved 21 October 2015. See also free online access version.
60. Markoff, Jack (21 October 2015). "Sorry, Einstein. Quantum Study Suggests 'Spooky Action' Is Real.". New York Times. Retrieved 21 October 2015.
61. Arndt, M; Nairz, O; Vos-Andreae, J; Keller, C; van der Zouw, G; Zeilinger, A (14 October 1999). "Wave–particle duality of C60molecules". Nature. 401: 680–682. Bibcode:1999Natur.401..680A. doi:10.1038/44348. PMID 18494170. (subscription required)
62. Olaf Nairz, Markus Arndt, and Anton Zeilinger, "Quantum interference experiments with large molecules", American Journal of Physics, 71 (April 2003) 319–325.

63. Lee, K. C.; Sprague, M. R.; Sussman, B. J.; Nunn, J.; Langford, N. K.; Jin, X.- M.; Champion, T.; Michelberger, P.; Reim, K. F.; England, D.; Jaksch, D.; Walmsley, I. A. (2 December 2011). "Entangling macroscopic diamonds at room temperature". Science. 334 (6060): 1253–1256. Bibcode:2011Sci...334.1253L. doi:10.1126/science.1211914. PMID 22144620. Lay summary.
64. sciencemag.org, supplementary materials
65. Kumar, M., Quantum, Icon Books, 2009, p. 313.
66. Alisa Bokulich, Gregg Jaeger, Philosophy of Quantum Information and Entanglement, Cambridge University Press, 2010, xv.
67. Letter from Einstein to Max Born, 3 March 1947; The Born-Einstein Letters; Correspondence between Albert Einstein and Max and Hedwig Born from 1916 to 1955, Walker, New York, 1971. (cited in M. P. Hobson; et al., Quantum Entanglement and Communication Complexity (1998), pp. 1/13, CiteSeerX 10.1.1.20.8324)
68. J. S. Bell (1964). "On the Einstein-Poldolsky-Rosen paradox". Physics.
69. Freedman, Stuart J.; Clauser, John F. (1972). "Experimental Test of Local Hidden-Variable Theories". Physical Review Letters. 28 (14): 938–941. Bibcode:1972PhRvL..28..938F. doi:10.1103/PhysRevLett.28.938.
70. A. Aspect; P. Grangier & G. Roger (1982). "Experimental Realization of Einstein-Podolsky-Rosen-Bohm Gedankenexperiment: A New Violation of Bell's Inequalities". Physical Review Letters. 49 (2): 91–94. Bibcode:1982PhRvL..49...91A. doi:10.1103/PhysRevLett.49.91.
71. Hanson, Ronald. "Loophole-free Bell inequality violation using electron spins separated by 1.3 kilometres". Nature. 526: 682–686. arXiv:1508.05949 . Bibcode:2015Natur.526..682H. doi:10.1038/nature15759.
72. C. H. Bennett and G. Brassard. "Quantum cryptography: Public key distribution and coin tossing". In Proceedings of IEEE International Conference on Computers, Systems and Signal Processing, volume 175, page 8. New York, 1984. http://researcher.watson.ibm.com/researcher/files/us-bennetc/BB84highest.pdf
73. Ekert, A.K. (1991). "Quantum cryptography based on Bell's theorem". Phys. Rev. Lett. APS. 67 (6): 661–663. Bibcode:1991PhRvL..67..661E. doi:10.1103/PhysRevLett.67.661. ISSN 0031-9007. PMID 10044956. (subscription required (help)).
74. Asher Peres, Quantum Theory, Concepts and Methods, Kluwer, 1993; ISBN 0-7923-2549-4 p. 115.
75. Rupert W., Anderson (28 March 2015). The Cosmic Compendium: Interstellar Travel (First ed.). The Cosmic Compendium. p. 100. ISBN 9781329022027.
76. I. Gerhardt; Q. Liu; A. Lamas-Linares; J. Skaar; V. Scarani; V. Makarov; C. Kurtsiefer (2011), "Experimentally faking the violation of Bell's inequalities", Phys. Rev. Lett., 107 (17): 170404, arXiv:1106.3224 , Bibcode:2011PhRvL.107q0404G, doi:10.1103/PhysRevLett.107.170404
77. Santos, E (2004). "The failure to perform a loophole-free test of Bell's

Inequality supports local realism". Foundations of Physics. 34: 1643–1673. Bibcode:2004FoPh...34.1643S. doi:10.1007/s10701-004-1308-z.

78. H. Zbinden; et al. (2001). "Experimental test of nonlocal quantum correlations in relativistic configurations". Phys. Rev. A. 63 (2): 22111. arXiv:quant-ph/0007009 . Bibcode:2001PhRvA..63b2111Z. doi:10.1103/PhysRevA.63.022111.
79. Some of the history of both referenced Zbinden, et al. experiments is provided in Gilder, L., The Age of Entanglement, Vintage Books, 2008, pp. 321–324.
80. Cirel'son, B. S. (1980). "Quantum generalizations of Bell's inequality". Letters in Mathematical Physics. 4 (2): 93–100. Bibcode:1980LMaPh...4...93C. doi:10.1007/BF00417500.
81. Xiao-song Ma, Stefan Zotter, Johannes Kofler, Rupert Ursin, Thomas Jennewein, Časlav Brukner & Anton Zeilinger; Zotter; Kofler; Ursin; Jennewein; Brukner; Zeilinger (26 April 2012). "Experimental delayed-choice entanglement swapping". Nature Physics. 8 (6): 480–485. arXiv:1203.4834 . Bibcode:2012NatPh...8..480M. doi:10.1038/nphys2294.
82. Megidish, E.; Halevy, A.; Shacham, T.; Dvir, T.; Dovrat, L.; Eisenberg, H. S. (2013). "Entanglement Swapping between Photons that have Never Coexisted". Physical Review Letters. 110 (21). arXiv:1209.4191 . Bibcode:2013PhRvL.110u0403M. doi:10.1103/physrevlett.110.210403.
83. "Classical carrier could create entanglement". physicsworld.com. Retrieved 2014-06-14.
84. http://hansonlab.tudelft.nl/loophole-free-bell-test/
85. "Entangled photons make a picture from a paradox". Nature News & Comment. Retrieved 13 October 2014.
86. "Time from quantum entanglement: an experimental illustration". Physical Review A. 89. arXiv:1310.4691 . Bibcode:2014PhRvA..89e2122M. doi:10.1103/PhysRevA.89.052122.
87. "Entangled toy universe shows time may be an illusion". Retrieved 13 October 2014.
88. David Deutsch, The Beginning of infinity. Page 299
89. "Quantum Experiment Shows How Time ‹Emerges' from Entanglement". Medium. Retrieved 13 October 2014.
90. "New Quantum Theory Could Explain the Flow of Time". WIRED. Retrieved 13 October 2014.
91. Werner, R.F. (1989). "Quantum States with Einstein-Podolsky-Rosen correlations admitting a hidden-variable model". Physical Review A. 40 (8): 4277–4281. Bibcode:1989PhRvA..40.4277W. doi:10.1103/PhysRevA.40.4277. PMID 9902666.
92. Jaeger G, Shimony A, Vaidman L; Shimony; Vaidman (1995). "Two Interferometric Complementarities". Phys. Rev. 51 (1): 54–67. Bibcode:1995PhRvA..51...54J. doi:10.1103/PhysRevA.51.54.
93. Nielsen, Michael A.; Chuang, Isaac L. (2000). Quantum Computation and Quantum Information. Cambridge University Press. pp. 112–113. ISBN 0-521-

63503-9.

94. Laloe, Franck (2012), Do We Really Understand Quantum Mechanics, Cambridge University Press, ISBN 978-1-107-02501-1

95. Gurvits L (2003). "Classical deterministic complexity of Edmonds' Problem and quantum entanglement". Proceedings of the thirty-fifth annual ACM symposium on Theory of computing: 10. doi:10.1145/780542.780545. ISBN 1-58113-674-9.

96. Horodecki M, Horodecki P, Horodecki R; Horodecki; Horodecki (1996). "Separability of mixed states: necessary and sufficient conditions". Physics Letters A. 223: 210. arXiv:quant-ph/9605038 . Bibcode:1996PhLA..223....1H. doi:10.1016/S0375-9601(96)00706-2.

97. Dirac, P. A. M. (2008). "Note on Exchange Phenomena in the Thomas Atom". Mathematical Proceedings of the Cambridge Philosophical Society. 26 (3): 376. Bibcode:1930PCPS...26..376D. doi:10.1017/S0305004100016108.

98. Fan, H; Korepin V; Roychowdhury V (26 November 2004). "Entanglement in a Valence-Bond Solid State". Physical Review Letters. 93 (22): 227203. arXiv:quant-ph/0406067 . Bibcode:2004PhRvL..93v7203F. doi:10.1103/PhysRevLett.93.227203. PMID 15601113.

99. Franchini, F.; Its, A. R.; Korepin, V. E.; Takhtajan, L. A. (2010). "Spectrum of the density matrix of a large block of spins of the XY model in one dimension". Quantum Information Processing. 10 (3): 325–341. arXiv:1002.2931 . doi:10.1007/s11128-010-0197-7.

100. Cerf, Nicolas J.; Cleve, Richard. "Information-theoretic interpretation of quantum error-correcting codes" (PDF).

101. Plenio; Virmani (2007). "An introduction to entanglement measures". Quant. Inf. Comp. 1: 1–51. arXiv:quant-ph/0504163 . Bibcode:2005quant.ph..4163P.

102. Wolchover, Natalie (25 April 2014). "New Quantum Theory Could Explain the Flow of Time". www.wired.com. Quanta Magazine. Retrieved 27 April 2014.

103. Huang, Yichen (21 March 2014). "Computing quantum discord is NP-complete". New Journal of Physics. 16 (3): 033027. arXiv:1305.5941 . Bibcode:2014NJPh...16c3027H. doi:10.1088/1367-2630/16/3/033027.

104. Bouwmeester, Dik; Pan, Jian-Wei; Mattle, Klaus; Eibl, Manfred; Weinfurter, Harald & Zeilinger, Anton (1997). "Experimental Quantum Teleportation" (PDF). Nature. 390 (6660): 575–579. Bibcode:1997Natur.390..575B. doi:10.1038/37539.

105. Richard Jozsa; Noah Linden (2002). "On the role of entanglement in quantum computational speed-up". Proceedings of the Royal Society A: Mathematical, Physical and Engineering Sciences. 459 (2036): 2011–2032. arXiv:quant-ph/0201143 . Bibcode:2003RSPSA.459.2011J. doi:10.1098/rspa.2002.1097.

106. Ekert, Artur K. (1991). "Quantum cryptography based on Bell's theorem". Physical Review Letters. 67 (6): 661–663. Bibcode:1991PhRvL..67..661E. doi:10.1103/PhysRevLett.67.661. PMID 10044956.

107. Karol Horodecki; Michal Horodecki; Pawel Horodecki; Ryszard Horodecki; Marcin Pawlowski; Mohamed Bourennane (2010). "Contextuality offers device-

independent security". arXiv:1006.0468 quant-ph.

108. Pezze, Luca & Smerzi, Augusto (2009). "Entanglement, Nonlinear Dynamics, and the Heisenberg Limit" (PDF). Phys. Rev. Lett. 102 (10): 100401. arXiv:0711.4840 . Bibcode:2009PhRvL.102j0401P. doi:10.1103/PhysRevLett.102.100401. PMID 19392092.
109. Database error – Qwiki
110. Kitagawa, Masahiro; Ueda, Masahito (1993). "Squeezed Spin States". Phys. Rev. A. 47: 5138–5143. Bibcode:1993PhRvA..47.5138K. doi:10.1103/physreva.47.5138.
111. Wineland, D. J.; Bollinger, J. J.; Itano, W. M.; Moore, F. L.; Heinzen, D. J. "Spin squeezing and reduced quantum noise in spectroscopy". Phys. Rev. A. 46: R6797–R6800. Bibcode:1992PhRvA..46.6797W. doi:10.1103/PhysRevA.46.R6797.
112. "Phys. Rev. Lett. 71, 1355 (1993): Interferometric detection of optical phase shifts at the Heisenberg limit". Physical Review Letters. Retrieved 13 October 2014.
113. Horodecki R, Horodecki P, Horodecki M, Horodecki K; Horodecki; Horodecki; Horodecki (2007). "Quantum entanglement". Rev. Mod. Phys. 81 (2): 865–942. arXiv:quant-ph/0702225 . Bibcode:2009RvMP...81..865H. doi:10.1103/RevModPhys.81.865.
114. Gurvits, L., Classical deterministic complexity of Edmonds' problem and quantum entanglement, in Proceedings of the 35th ACM Symposium on Theory of Computing, ACM Press, New York, 2003.
115. Sevag Gharibian, Strong NP-Hardness of the Quantum Separability Problem, Quantum Information and Computation, Vol. 10, No. 3&4, pp. 343–360, 2010. arXiv:0810.4507.
116. "Geometrical aspects of entanglement", Physical Review A 74, 012313 (2006)
117. http://physicsworld.com/cws/article/news/2016/aug/16/china-launches-world-s-first-quantum-science-satellite
118. Frank Jensen: Introduction to Computational Chemistry. Wiley, 2007, ISBN 978-0-470-01187-4.
119. Berkeley Lab Press Release: Untangling the Quantum Entanglement Behind Photosynthesis: Berkeley scientists shine new light on green plant secrets.
120. Mohan Sarovar, Akihito Ishizaki, Graham R. Fleming, K. Birgitta Whaley: Quantum entanglement in photosynthetic light harvesting complexes. arXiv:0905.3787
121. Aspect, Alain; Dalibard, Jean; Roger, Gérard (December 1982). "Experimental Test of Bell's Inequalities Using Time- Varying Analyzers". Physical Review Letters. 49 (25): 1804–1807. Bibcode:1982PhRvL..49.1804A. doi:10.1103/PhysRevLett.49.1804.
122. Rowe MA, et al. (February 2001). "Experimental violation of a Bell's Inequality with efficient detection". Nature. 409 (6822): 791–794. Bibcode:2001Natur.409..791K. doi:10.1038/35057215.
123. Gilles Brassard, Richard Cleve, Alain Tapp, "The cost of exactly simulating

quantum entanglement with classical communication".

124. Ghirardi, G.C.; Rimini, A.; Weber, T. (March 1980). "A general argument against superluminal transmission through the quantum mechanical measurement process". Lettere Al Nuovo Cimento. 27 (10): 293–298. doi:10.1007/BF02817189.
125. Popescu, Sandu; Rohrlich, Daniel (1994). "Nonlocality as an axiom". Foundations of Physics. 24 (3): 379–385. Bibcode:1994FoPh...24..379P. doi:10.1007/BF02058098.
126. Einstein, Albert; Podolsky, Boris; Rosen, Nathan (May 1935). "Can Quantum-Mechanical Description of Physical Reality Be Considered Complete?". Physical Review. 47 (10): 777–780. Bibcode:1935PhRv...47..777E. doi:10.1103/PhysRev.47.777.
127. Bohm, D. (1951). Quantum Theory, Prentice-Hall, Englewood Cliffs, page 29, and Chapter 5 section 3, and Chapter 22 Section 19.
128. Nielsen, Michael A.; Chuang, Isaac L. (2000). Quantum Computation and Quantum Information. Cambridge University Press. pp. 112–113. ISBN 0-521-63503-9.
129. Bohr, N (July 1935). "Can Quantum-Mechanical Description of Physical Reality Be Considered Complete?". Physical Review. 48 (8): 696. Bibcode:1935PhRv...48..696B. doi:10.1103/PhysRev.48.696.
130. Furry, W.H. (March 1936). "Remarks on Measurements in Quantum Theory". Physical Review. 49 (6): 476. Bibcode:1936PhRv...49..476F. doi:10.1103/PhysRev.49.476.
131. von Neumann, J. (1932/1955). In Mathematische Grundlagen der Quantenmechanik, Springer, Berlin, translated into English by Beyer, R.T., Princeton University Press, Princeton, cited by Baggott, J. (2004) Beyond Measure: Modern physics, philosophy, and the meaning of quantum theory, Oxford University Press, Oxford, ISBN 0-19-852927-9, pages 144-145.
132. Maudlin, T., 1994, Quantum Non-Locality and Relativity: Metaphysical Intimations of Modern Physics, Cambridge, Massachusetts: Blackwell.
133. "Quantum Mechanics and Reality" ("Quanten-Mechanik und Wirklichkeit", Dialectica 2:320-324, 1948)
134. Bell, John (1964). "On the Einstein Podolsky Rosen paradox". Physics. 1: 195.
135. Clauser, John F.; Horne, Michael A.; Shimony, Abner; Holt, Richard A. (October 1969). "Proposed Experiment to Test Local Hidden-Variable Theories". Physical Review Letters. 23 (15): 880–884. Bibcode:1969PhRvL..23..880C. doi:10.1103/PhysRevLett.23.880.
136. Hardy, Lucien (1993). "Nonlocality for two particles without inequalities for almost all entangled states". Physical Review Letters. 71 (11): 1665–1668. Bibcode:1993PhRvL..71.1665H. doi:10.1103/PhysRevLett.71.1665.
137. Cirel'son, B. S. (1980). "Quantum generalizations of Bell's inequality". Letters in Mathematical Physics. 4 (2): 93–100. Bibcode:1980LMaPh...4...93C. doi:10.1007/BF00417500.

138. Barrett, J.; Linden, N.; Massar, S.; Pironio, S.; Popescu, S.; Roberts, D. (2005). "Non-local correlations as an information theoretic resource". Physical Review A. 71 (2): 022101. arXiv:quant-ph/0404097 . Bibcode:2005PhRvA..71b2101B. doi:10.1103/PhysRevA.71.022101.

139. Werner, R.F. (1989). "Quantum States with Einstein-Podolsky-Rosen correlations admitting a hidden-variable model". Physical Review A. 40 (8): 4277–4281. Bibcode:1989PhRvA..40.4277W. doi:10.1103/PhysRevA.40.4277. PMID 9902666.

140. Barrett, Jonathan; Pironio, Stefano (September 2005). "Popescu-Rohrlich Correlations as a Unit of Nonlocality". Physical Review Letters. 95 (14): 140401. arXiv:quant-ph/0506180 . Bibcode:2005PhRvL..95n0401B. doi:10.1103/PhysRevLett.95.140401. PMID 16241631.

141. Michael M. Wolf; David Perez-Garcia; Carlos Fernandez (2009). "Measurements Incompatible in Quantum Theory Cannot Be Measured Jointly in Any Other No-Signaling Theory". Science. 330 (23): 230402. arXiv:0905.2998 . Bibcode:2009PhRvL.103w0402W. doi:10.1103/PhysRevLett.103.230402.

142. Jonathan Oppenheim; Stephanie Wehner (2010). "The uncertainty principle determines the non-locality of quantum mechanics". Science. 330 (6007): 1072–1074. arXiv:1004.2507 . Bibcode:2010Sci...330.1072O. doi:10.1126/science.1192065. PMID 21097930.

143. van Dam, Wim (2005). "Implausible Consequences of Superstrong Nonlocality". arXiv:quant-ph/0501159 .

144. Pawlowski, M.; Paterek, T.; Kaszlikowski, D.; Scarani, V.; Winter, A.; Zukowski, M. (October 2009). "Information Causality as a Physical Principle". Nature. 461 (7267): 1101–1104. arXiv:0905.2292 . Bibcode:2009Natur.461.1101P. doi:10.1038/nature08400.

145. http://arxiv.org/pdf/1210.0194.pdf - If no information gain implies no disturbance, then any discrete physical theory is classical

146. Barrett, Jonathan; Hardy, Lucien; Kent, Adrian (2005). "No Signalling and Quantum Key Distribution". Physical Review Letters. 95 (1): 010503. arXiv:quant-ph/0405101 . Bibcode:2005PhRvL..95a0503B. doi:10.1103/PhysRevLett.95.010503. PMID 16090597.

147. Bennett, C.H.; DiVincenzo, D.P.; Fuchs, C.A.; Mor, T.; Rains, E.; Shor, P.W.; Smolin, J.A.; Wootters, W.K. (1999). "Quantum nonlocality without entanglement". Physical Review A. 59 (2): 1070—1091. arXiv:quant-ph/9804053 . doi:10.1103/PhysRevA.59.1070.

148. Marius Junge; Carlos Palazuelos (2010). "Large violation of Bell inequalities with low entanglement". Communications in Mathematical Physics. 306 (3): 695–746. arXiv:1007.3043v2 . Bibcode:2011CMaPh.306..695J. doi:10.1007/s00220-011-1296-8.

149. Thomas Vidick; Stephanie Wehner (2010). "More Non-locality with less Entanglement". Physical Review A. 83 (5). arXiv:1011.5206v2 . Bibcode:2011PhRvA..83e2310V. doi:10.1103/PhysRevA.83.052310.

150. Yeong-Cherng Liang; Tamas Vertesi; Nicolas Brunner (2010). "Semi-

device-independent bounds on entanglement”. Physical Review A. 83 (2). arXiv:1012.1513v2 . Bibcode:2011PhRvA..83b2108L. doi:10.1103/PhysRevA.83.022108.

151. Hermann Wimmel (1992). Quantum physics & observed reality: a critical interpretation of quantum mechanics. World Scientific. p. 2. ISBN 978-981-02-1010-6. Retrieved 9 May 2011.
152. J. J. Sakurai (1994). Modern Quantum Mechanics (2nd ed.). ISBN 0201539292.
153. George S. Greenstein & Arthur G. Zajonc (2006). The Quantum Challenge: Modern Research On The Foundations Of Quantum Mechanics (2nd ed.). ISBN 076372470X.
154. Wojciech H. Zurek, Decoherence, einselection, and the quantum origins of the classical,Reviews of Modern Physics 2003, 75, 715 or http://arxiv.org/abs/quant-ph/0105127
155. M.O. Scully; W.E. Lamb; A. Barut (1987). “On the theory of the Stern–Gerlach apparatus” (PDF). Foundations of Physics. 17: 575–583. Bibcode:1987FoPh...17..575S. doi:10.1007/BF01882788. Retrieved 9 November 2012.
156. C. Misner; K. Thorne; W. Zurek (2009). “John Wheeler, relativity and quantum information” (PDF). Physics Today. 64: 40.
157. Y.Y. Huang (2016). “Classical harmonic vibrations with micro amplitudes and low frequencies monitored by quantum entanglement”. Optic Review. 23: 92. doi:10.1007/s10043-015-0151-0.
158. Y.Y. Huang (2014). “One atomic beam as a detector of classical harmonic vibrations with micro amplitudes and low frequencies”. Journal of the Korean Physical Society. 64: 775. doi:10.3938/jkps.64.775.
159. Y.Y. Huang (2014). “Detecting the classical harmonic vibrations of micro amplitudes and low frequencies with an atomic Mach–Zehnder interferometer”. General Relativity and Gravitation. 46: 1614. doi:10.1007/s10714-013-1614-x.
160. Hrvoje Nikolić (2007). “Quantum mechanics: Myths and facts”. Foundation of Physics. 37: 1563–1611. arXiv:quant-ph/0609163 . Bibcode:2007FoPh...37.1563N. doi:10.1007/s10701-007-9176-y.
161. S. Gröblacher; et al. (2007). “An experimental test of non-local realism”. Nature. 446 (871): 871–5. arXiv:0704.2529 . Bibcode:2007Natur.446..871G. doi:10.1038/nature05677. PMID 17443179.
162.

CHAPTER 4

Formulations in Quantum Mechanics

Chapter Contents

- **Introduction**
- **Heisenberg Picture**
- **Interaction Picture**
- **Matrix Mechanics**

Different formulations in both classical and quantum mechanics provide exactly the same predictions and experimental results. But the mathematics of one formulation may yield solutions of a specific physical situation much more quickly and easily than others. Some are more easily generalized to relativistic mechanics. Some help us to "visualize" what is going on - the so-called "elements of reality" - in particular problems. Some are better at eigenvalues than transition probabilities, some better for bound/periodic systems than for "free" particles. Some find "constants of the motion" better. Some are better for quantum field theory than for quantum electrodynamics and so on.

INTRODUCTION

The mathematical formulations of quantum mechanics are those mathematical formalisms that permit a rigorous description of quantum mechanics. Such are distinguished from mathematical formalisms for theories developed prior to the early 1900s by the use of abstract mathematical structures, such as infinite-dimensional Hilbert spaces and operators on these spaces. Many of these structures are drawn from functional analysis, a research area within pure mathematics that was influenced in part by the needs of quantum mechanics.

In brief, values of physical observables such as energy and momentum were no longer considered as values of functions on phase space, but as eigenvalues; more precisely as spectral values of linear operators in Hilbert space.

These formulations of quantum mechanics continue to be used today. At the heart of the description are ideas of quantum state and quantum observable which are radically different from those used in previous models of physical reality. While the mathematics permits calculation of many quantities that can be measured experimentally, there is a definite theoretical limit to values that can be simultaneously measured. This limitation was first elucidated by Heisenberg through a thought experiment, and is represented mathematically in the new formalism by the non-commutativity of operators representing quantum observables.

Prior to the emergence of quantum mechanics as a separate theory, the mathematics used in physics consisted mainly of formal mathematical analysis, beginning with calculus, and increasing in complexity up to differential geometry and partial differential equations. Probability theory was used in statistical mechanics. Geometric intuition played a strong role in the first two and, accordingly, theories of relativity were formulated entirely in terms of geometric concepts. The phenomenology of quantum physics arose roughly between 1895 and 1915, and for the 10 to 15 years before the emergence of quantum theory (around 1925) physicists continued to think of quantum theory within the confines of what is now called classical physics, and in particular within the same mathematical structures. The most sophisticated example of this is the Sommerfeld–Wilson–Ishiwara quantization rule, which was formulated entirely on the classical phase space.

The "Old Quantum Theory" and the Need for New Mathematics

In the 1890s, Planck was able to derive the blackbody spectrum which was later used to avoid the classical ultraviolet catastrophe by making the unorthodox assumption that, in the interaction of electromagnetic radiation with matter, energy could only be exchanged in discrete units which he called quanta. Planck postulated a direct proportionality between the frequency of radiation and the quantum of energy at that frequency. The proportionality constant, h, is now called Planck's constant in his honor.

In 1905, Einstein explained certain features of the photoelectric effect by assuming that Planck's energy quanta were actual particles, which were later dubbed photons.

All of these developments were phenomenological and challenged the theoretical physics of the time. Bohr and Sommerfeld went on to modify classical mechanics in an attempt to deduce the Bohr model from first principles. They proposed that, of all closed classical orbits traced by a mechanical system in its phase space, only the ones that enclosed an area which was a multiple of Planck's constant were actually allowed. The most sophisticated version of this formalism was the so-called Sommerfeld–Wilson–Ishiwara quantization. Although the Bohr model of the hydrogen atom could be explained in this way, the spectrum of the helium atom (classically an unsolvable 3-body problem) could not be predicted. The mathematical status of quantum theory remained uncertain for some time.

In 1923 de Broglie proposed that wave–particle duality applied not only to photons but to electrons and every other physical system.

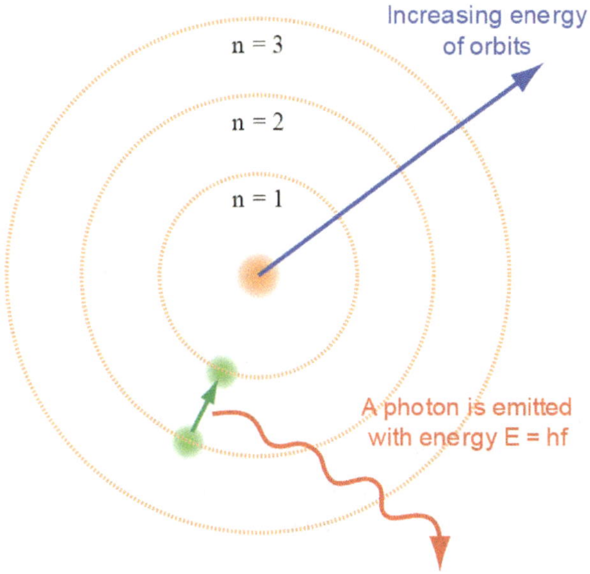

The situation changed rapidly in the years 1925–1930, when working mathematical foundations were found through the groundbreaking work of Erwin Schrödinger, Werner Heisenberg, Max Born, Pascual Jordan, and the foundational work of John von Neumann, Hermann Weyl and Paul Dirac, and it became possible to unify several different approaches in terms of a fresh set of ideas. The physical interpretation of the theory was also clarified in these years after Werner Heisenberg discovered the uncertainty relations and Niels Bohr introduced the idea of complementarity.

The "New Quantum Theory"

Werner Heisenberg's matrix mechanics was the first successful attempt at replicating the observed quantization of atomic spectra. Later in the same year, Schrödinger created his wave mechanics. Schrödinger's formalism was considered easier to understand, visualize and calculate as it led to differential equations, which physicists were already familiar with solving. Within a year, it was shown that the two theories were equivalent.

Schrödinger himself initially did not understand the fundamental probabilistic nature of quantum mechanics, as he thought that the absolute square of the wave function of an electron should be interpreted as the charge density of an object smeared out over an extended, possibly infinite, volume of space. It was Max Born who introduced the interpretation of the absolute square of the wave function as the probability distribution of the position of a pointlike object. Born's idea was soon taken over by Niels Bohr in Copenhagen who then became the father of the Copenhagen interpretation of quantum mechanics. Schrödinger's wave function can be seen to be closely related to the classical Hamilton–Jacobi equation. The correspondence to classical mechanics was even more explicit, although somewhat more formal, in Heisenberg's matrix mechanics. In his PhD thesis project, Paul Dirac discovered that the equation for the operators in the Heisenberg representation, as it is now called, closely translates to classical equations for the dynamics of certain quantities in the Hamiltonian formalism of classical mechanics, when one expresses them through Poisson brackets, a procedure now known as canonical quantization.

To be more precise, already before Schrödinger, the young postdoctoral fellow Werner Heisenberg invented his matrix mechanics, which was the first correct quantum mechanics— the essential breakthrough. Heisenberg's matrix mechanics formulation was based on algebras of infinite matrices, a very radical formulation in light of the mathematics of classical physics, although he started from the index-terminology of the experimentalists of that time, not even aware that his index-schemes were matrices, as Born soon pointed out to him. In fact, in these early years, linear algebra was not generally popular with physicists in its present form.

Although Schrödinger himself after a year proved the equivalence of his wave-mechanics and Heisenberg's matrix mechanics, the reconciliation of the two approaches and their modern abstraction as motions in Hilbert space is generally attributed to Paul Dirac, who wrote a lucid account in his 1930 classic The Principles of Quantum Mechanics. He is the third, and possibly most important, pillar of that field (he soon was the only one to have discovered a relativistic generalization of the theory). In his above-mentioned account, he introduced the bra–ket notation, together with an abstract formulation in terms of the Hilbert space used in functional analysis; he showed that Schrödinger's and Heisenberg's approaches were two different representations of the same theory, and found a third, most general one, which represented the dynamics of the system. His work was particularly fruitful in all kinds of generalizations of the field.

The first complete mathematical formulation of this approach, known as the Dirac–von Neumann axioms, is generally credited to John von Neumann's 1932 book Mathematical Foundations of Quantum Mechanics, although Hermann Weyl had already referred to Hilbert spaces (which he called unitary spaces) in his 1927 classic paper and book. It was developed in parallel with a new approach to the mathematical spectral theory based on linear operators rather than the quadratic forms that were David Hilbert's approach a generation earlier. Though theories of quantum mechanics continue to evolve to this day, there is a basic framework for the mathematical formulation of quantum mechanics which underlies most approaches and can be traced back to the mathematical work of John von Neumann. In other words, discussions about interpretation of the theory, and extensions to it, are now mostly conducted on the basis of shared assumptions about the mathematical foundations.

Later Developments

The application of the new quantum theory to electromagnetism resulted in quantum field theory, which was developed starting around 1930. Quantum field theory has driven the development of more sophisticated formulations of quantum mechanics, of which the one presented here is a simple special case.

- Path integral formulation
- Phase space formulation of Quantum Mechanics & geometric quantization
- Signed Particle Formulation
- quantum field theory in curved spacetime
- axiomatic, algebraic and constructive quantum field theory

- C* algebra formalism
- Generalized Statistical Model of Quantum Mechanics

On a different front, von Neumann originally dispatched quantum measurement with his infamous postulate on the collapse of the wavefunction, raising a host of philosophical problems. Over the intervening 70 years, the problem of measurement became an active research area and itself spawned some new formulations of quantum mechanics.

- Relative state/Many-worlds interpretation of quantum mechanics
- Decoherence
- Consistent histories formulation of quantum mechanics
- Quantum logic formulation of quantum mechanics

A related topic is the relationship to classical mechanics. Any new physical theory is supposed to reduce to successful old theories in some approximation. For quantum mechanics, this translates into the need to study the so-called classical limit of quantum mechanics. Also, as Bohr emphasized, human cognitive abilities and language are inextricably linked to the classical realm, and so classical descriptions are intuitively more accessible than quantum ones. In particular, quantization, namely the construction of a quantum theory whose classical limit is a given and known classical theory, becomes an important area of quantum physics in itself.

Finally, some of the originators of quantum theory (notably Einstein and Schrödinger) were unhappy with what they thought were the philosophical implications of quantum mechanics. In particular, Einstein took the position that quantum mechanics must be incomplete, which motivated research into so-called hidden-variable theories. The issue of hidden variables has become in part an experimental issue with the help of quantum optics.

- de Broglie–Bohm–Bell pilot wave formulation of quantum mechanics
- Bell's inequalities
- Kochen–Specker theorem

Mathematical Structure of Quantum Mechanics

A physical system is generally described by three basic ingredients: states; observables; and dynamics (or law of time evolution) or, more generally, a group of physical symmetries. A classical description can be given in a fairly direct way by a phase space model of mechanics: states are points in a symplectic phase space, observables are real-valued functions on it, time evolution is given by a one-parameter group of symplectic transformations of the phase space, and physical symmetries are realized by symplectic transformations. A quantum description consists of a Hilbert space of states, observables are self adjoint operators on the space of states, time evolution is given by a one-parameter group of unitary transformations on the Hilbert space of states, and physical symmetries are realized by unitary transformations.

Postulates of Quantum Mechanics

The following summary of the mathematical framework of quantum mechanics can be partly traced back to the Dirac–von Neumann axioms.

- Each physical system is associated with a (topologically) separable complex Hilbert space H with inner product . Rays (one-dimensional subspaces) in H are associated with states of the system. In other words, physical states can be identified with equivalence classes of vectors of length 1 in H, where two vectors represent the same state if they differ only by a phase factor. Separability is a mathematically convenient hypothesis, with the physical interpretation that countably many observations are enough to uniquely determine the state. "A quantum mechanical state is a ray in projective Hilbert space, not a vector. Many textbooks fail to make this distinction, which could be partly a result of the fact that the Schrödinger equation itself involves Hilbert-space "vectors", with the result that the imprecise use of "state vector" rather than ray is very difficult to avoid."
- The Hilbert space of a composite system is the Hilbert space tensor product of the state spaces associated with the component systems (for instance, J.M. Jauch, Foundations of quantum mechanics, section 11-7). For a non-relativistic system consisting of a finite number of distinguishable particles, the component systems are the individual particles.
- Physical symmetries act on the Hilbert space of quantum states unitarily or antiunitarily due to Wigner's theorem (supersymmetry is another matter entirely).
- Physical observables are represented by Hermitian matrices on H.

The expectation value (in the sense of probability theory) of the observable A for the system in state represented by the unit vector $\psi \in$ H is

$$\langle \psi \mid A \mid \psi \rangle$$

By spectral theory, we can associate a probability measure to the values of A in any state ψ. We can also show that the possible values of the observable A in any state must belong to the spectrum of A. In the special case A has only discrete spectrum, the possible outcomes of measuring A are its eigenvalues. More precisely, if we represent the state ψ in the basis formed by the eigenvectors of A, then the square of the modulus of the component attached to a given eigenvector is the probability of observing its corresponding eigenvalue.

More generally, a state can be represented by a so-called density operator, which is a trace class, nonnegative self-adjoint operator ρnormalized to be of trace 1. The expected value of A in the state ρ is

$$\operatorname{tr}(A\rho)$$

If $\rho\psi$ is the orthogonal projector onto the one-dimensional subspace of H spanned by $|\psi\rangle$, then

Density operators are those that are in the closure of the convex hull of the one-dimensional orthogonal projectors. Conversely, one-dimensional orthogonal projectors are extreme points of the set of density operators. Physicists also call one-dimensional orthogonal projectors pure states and other density operators mixed states.

One can in this formalism state Heisenberg's uncertainty principle and prove it as a theorem, although the exact historical sequence of events, concerning who derived what and under which framework, is the subject of historical investigations outside the scope of this article.

Furthermore, to the postulates of quantum mechanics one should also add basic statements on the properties of spin and Pauli's exclusion principle, see below.

Pictures of Dynamics

- In the so-called Schrödinger picture of quantum mechanics, the dynamics is given as follows:

The time evolution of the state is given by a differentiable function from the real numbers **R**, representing instants of time, to the Hilbert space of system states. This map is characterized by a differential equation as follows: If $|\psi(t)\rangle$ denotes the state of the system at any one time t, the following Schrödinger equation holds:

Schrödinger equation *(general)*

$$i\hbar\frac{d}{dt}|\psi(t)\rangle = H|\psi(t)\rangle$$

where H is a densely defined self-adjoint operator, called the system Hamiltonian, i is the imaginary unit and ħ is the reduced Planck constant. As an observable, H corresponds to the total energy of the system.

Alternatively, by Stone's theorem one can state that there is a strongly continuous one-parameter unitary group U(t): H → H such that

$$|\psi(t+s)\rangle = U(t)|\psi(s)\rangle$$

for all times s, t. The existence of a self-adjoint Hamiltonian H such that

$$U(t) = e^{-(i/\hbar)tH}$$

is a consequence of Stone's theorem on one-parameter unitary groups. It is assumed that H does not depend on time and that the perturbation starts at $t_0 = 0$; otherwise one must use the Dyson series, formally written as

$$U(t) = \mathcal{T}\left[\exp\left(-\frac{i}{\hbar}\int_{t_0}^{t} \mathrm{d}t'\, H(t')\right)\right],$$

where T time-ordering symbol.

(This symbol permutes a product of noncommuting operators of the form

$$B_1(t_1)\cdot B_2(t_2)\cdot\cdots\cdot B_n(t_n)$$

into the uniquely determined re-ordered expression

$B_{i_1}(t_{i_1}) \cdot B_{i_2}(t_{i_2}) \cdot \cdots \cdot B_{i_n}(t_{i_n})$ with $t_{i_1} \geq t_{i_2} \geq \cdots \geq t_{i_n}$.

The result is a causal chain, the primary cause in the past on the utmost r.h.s., and finally the present effect on the utmost l.h.s. .)

- The Heisenberg picture of quantum mechanics focuses on observables and instead of considering states as varying in time, it regards the states as fixed and the observables as changing. To go from the Schrödinger to the Heisenberg picture one needs to define time-independent states and time-dependent operators thus:

$$|\psi\rangle = |\psi(0)\rangle$$
$$A(t) = U(-t)AU(t).$$

It is then easily checked that the expected values of all observables are the same in both pictures

$$\langle\psi \mid A(t) \mid \psi\rangle = \langle\psi(t) \mid A \mid \psi(t)\rangle$$

Heisenberg picture *(general)*

$$\frac{d}{dt}A(t) = \frac{i}{\hbar}[H, A(t)] + \frac{\partial A(t)}{\partial t},$$

which is true for time-dependent A = A(t). Notice the commutator expression is purely formal when one of the operators is unbounded. One would specify a representation for the expression to make sense of it.

- The so-called Dirac picture or interaction picture has time-dependent states and observables, evolving with respect to different Hamiltonians. This picture is most useful when the evolution of the observables can be solved exactly, confining any complications to the evolution of the states. For this reason, the Hamiltonian for the observables is called "free Hamiltonian" and the Hamiltonian for the states is called "interaction Hamiltonian". In symbols:

Dirac picture

$$i\hbar\frac{d}{dt}|\psi(t)\rangle = H_{\rm int}(t)\,|\psi(t)\rangle$$

$$i\hbar\frac{d}{dt}A(t) = [A(t), H_0].$$

The interaction picture does not always exist, though. In interacting quantum field theories, Haag's theorem states that the interaction picture does not exist. This is because the Hamiltonian cannot be split into a free and an interacting part within a superselection sector. Moreover, even if in the Schrödinger picture the Hamiltonian does not depend on time, e.g. $H = H_0 + V$, in the interaction picture it does, at least, if V does not commute with H_0, since

$$H_{\rm int}(t) \equiv e^{(i/\hbar)tH_0}\, V\, e^{(-i/\hbar)tH_0}.$$

So the above-mentioned Dyson-series has to be used anyhow.

The Heisenberg picture is the closest to classical Hamiltonian mechanics (for example, the commutators appearing in the above equations directly translate into the classical Poisson brackets); but this is already rather "high-browed", and the Schrödinger picture is considered easiest to visualize and understand by most people, to judge from pedagogical accounts of quantum mechanics. The Dirac picture is the one used in perturbation theory, and is specially associated to quantum field theory and many-body physics.

Similar equations can be written for any one-parameter unitary group of symmetries of the physical system. Time would be replaced by a suitable coordinate parameterizing the unitary group (for instance, a rotation angle, or a translation distance) and the Hamiltonian would be replaced by the conserved quantity associated to the symmetry (for instance, angular or linear momentum).

Representations

The original form of the Schrödinger equation depends on choosing a particular representation of Heisenberg's canonical commutation relations. The Stone–von Neumann theorem dictates that all irreducible representations of the finite-dimensional Heisenberg commutation relations are unitarily equivalent. A systematic understanding of its consequences has led to the phase space formulation of quantum mechanics, which works in full phase space instead of Hilbert space, so then with a more intuitive link to the classical limit thereof. This picture also simplifies considerations of quantization, the deformation extension from classical to quantum mechanics.

The quantum harmonic oscillator is an exactly solvable system where the different representations are easily compared. There, apart from the Heisenberg, or Schrödinger (position or momentum), or phase-space representations, one also encounters the Fock (number) representation and the Segal–Bargmann (Fock-space or coherent state) representation (named after Irving Segal and Valentine Bargmann). All four are unitarily equivalent.

Time as an Operator

The framework presented so far singles out time as the parameter that everything depends on. It is possible to formulate mechanics in such a way that time becomes itself an observable associated to a self-adjoint operator. At the classical level, it is possible to arbitrarily parameterize the trajectories of particles in terms of an unphysical parameter s, and in that case the time t becomes an additional generalized coordinate of the physical system. At the quantum level, translations in s would be generated by a "Hamiltonian" $H - E$, where E is the energy operator and H is the "ordinary" Hamiltonian. However, since s is an unphysical parameter, physical states must be left invariant by "s-evolution", and so the physical state space is the kernel of $H - E$ (this requires the use of a rigged Hilbert space and a renormalization of the norm).

This is related to the quantization of constrained systems and quantization of gauge theories. It is also possible to formulate a quantum theory of events where time becomes an observable.

Spin

In addition to their other properties, all particles possess a quantity called spin, an intrinsic angular momentum. Despite the name, particles do not literally spin around an axis, and quantum mechanical spin has no correspondence in classical physics. In the position representation, a spinless wavefunction has position **r** and time t as continuous variables, $\psi = \psi(\mathbf{r}, t)$, for spin wavefunctions the spin is an additional discrete variable: $\psi = \psi(\mathbf{r}, t, \sigma)$, where σ takes the values;

$$\sigma = -S\hbar, -(S-1)\hbar, \ldots, 0, \ldots, +(S-1)\hbar, +S\hbar.$$

That is, the state of a single particle with spin S is represented by a $(2S+1)$-component spinor of complex-valued wave functions.

Two classes of particles with *very different* behaviour are bosons which have integer spin ($S = 0, 1, 2\ldots$), and fermions possessing half-integer spin ($S = \frac{1}{2}, \frac{3}{2}, \frac{5}{2}, \ldots$).

Pauli's Principle

The property of spin relates to another basic property concerning systems of N identical particles: Pauli's exclusion principle, which is a consequence of the following permutation behaviour of an N-particle wave function; again in the position representation one must postulate that for the transposition of any two of the N particles one always should have

Pauli principle

$$\psi(\ldots, \mathbf{r}_i, \sigma_i, \ldots, \mathbf{r}_j, \sigma_j, \ldots) = (-1)^{2S} \cdot \psi(\ldots, \mathbf{r}_j, \sigma_j, \ldots, \mathbf{r}_i, \sigma_i, \ldots)$$

i.e., on transposition of the arguments of any two particles the wavefunction should reproduce, apart from a prefactor $(-1)^{2S}$ which is +1 for bosons, but (−1) for fermions. Electrons are fermions with S = 1/2; quanta of light are bosons with S = 1. In nonrelativistic quantum mechanics all particles are either bosons or fermions; in relativistic quantum theories also supersymmetric theories exist, where a particle is a linear combination of a bosonic and a fermionic part. Only in dimension d = 2 can one construct entities where $(-1)^{2S}$ is replaced by an arbitrary complex number with magnitude 1, called anyons.

Although spin and the Pauli principle can only be derived from relativistic generalizations of quantum mechanics the properties mentioned in the last two paragraphs belong to the basic postulates already in the non-relativistic limit. Especially, many important properties in natural science, e.g. the periodic system of chemistry, are consequences of the two properties.

The Problem of Measurement

The picture given in the preceding paragraphs is sufficient for description of a completely isolated system. However, it fails to account for one of the main differences between quantum mechanics and classical mechanics, that is, the effects of measurement. The von Neumann description of quantum measurement of an observable A, when the system is prepared in a pure state ψ is the following (note, however, that von Neumann's description dates back to the 1930s and is based on experiments as performed during that time – more specifically the Compton–Simon experiment; it is not applicable to most present-day measurements within the quantum domain):

- Let A have spectral resolution

$$A = \int \lambda \, d\mathbf{E}_A(\lambda),$$

where EA is the resolution of the identity (also called projection-valued measure) associated to A. Then the probability of the measurement outcome lying in an interval B of R is $|E_A(B)\ \psi|2$. In other words, the probability is obtained by integrating the characteristic function of B against the countably additive measure

$$\langle \psi \mid \mathbf{E}_A \psi \rangle.$$

- If the measured value is contained in B, then immediately after the measurement, the system will be in the (generally non-normalized) state $E_A(B)\psi$. If the measured value does not lie in B, replace B by its complement for the above state.

For example, suppose the state space is the n-dimensional complex Hilbert space $\mathbf{C}^n$ and A is a Hermitian matrix with eigenvalues λ_i, with corresponding eigenvectors ψ_i. The projection-valued measure associated with A, E_A, is then

$$\mathbf{E}_A(B) = |\psi_i\rangle\langle\psi_i|,$$

where B is a Borel set containing only the single eigenvalue λ_i. If the system is prepared in state

$$|\psi\rangle$$

Then the probability of a measurement returning the value λ_i can be calculated by integrating the spectral measure

$$\langle \psi \mid \mathbf{E}_A \psi \rangle$$

over B_i. This gives trivially

$$\langle\psi|\psi_i\rangle\langle\psi_i \mid \psi\rangle = |\langle\psi \mid \psi_i\rangle|^2.$$

The characteristic property of the von Neumann measurement scheme is that repeating the same measurement will give the same results. This is also called the projection postulate.

A more general formulation replaces the projection-valued measure with a positive-operator valued measure (POVM). To illustrate, take again the finite-dimensional case. Here we would replace the rank-1 projections

$$|\psi_i\rangle\langle\psi_i|$$

by a finite set of positive operators

$$F_i F_i^*$$

whose sum is still the identity operator as before (the resolution of identity). Just as a set of possible outcomes $\{\lambda_1 \ldots \lambda_n\}$ is associated to a projection-valued measure, the same can be said for a POVM. Suppose the measurement outcome is λ_i. Instead of collapsing to the (unnormalized) state

$$|\psi_i\rangle\langle\psi_i|\psi\rangle$$

after the measurement, the system now will be in the state

$$F_i|\psi\rangle.$$

Since the $F_i F_i^*$ operators need not be mutually orthogonal projections, the projection postulate of von Neumann no longer holds.

The same formulation applies to general mixed states.

In von Neumann's approach, the state transformation due to measurement is distinct from that due to time evolution in several ways. For example, time evolution is deterministic and unitary whereas measurement is non-deterministic and non-unitary. However, since both types of state transformation take one quantum state to another, this difference was viewed by many as unsatisfactory. The POVM formalism views measurement as one among many other quantum operations, which are described by completely positive maps which do not increase the trace.

In any case it seems that the above-mentioned problems can only be resolved if the time evolution included not only the quantum system, but also, and essentially, the classical measurement apparatus (see above).

The Relative State Interpretation

An alternative interpretation of measurement is Everett's relative state interpretation, which was later dubbed the many-worlds interpretation of quantum physics.

List of Mathematical Tools

Part of the folklore of the subject concerns the mathematical physics textbook Methods of Mathematical Physics put together by Richard Courant from David Hilbert's Göttingen University courses. The story is told (by mathematicians) that physicists had dismissed the material as not interesting in the current research areas, until the advent of Schrödinger's equation. At that point it was realised that the mathematics of the new quantum mechanics was already laid out in it. It is also said that Heisenberg had consulted Hilbert about his matrix mechanics, and Hilbert observed that his own experience with infinite-dimensional matrices had derived from differential equations, advice which Heisenberg ignored, missing the opportunity to unify the theory as Weyl and Dirac did a few years later. Whatever the basis of the anecdotes, the mathematics of the theory was conventional at the time, whereas the physics was radically new.

The main tools include:

- linear algebra: complex numbers, eigenvectors, eigenvalues
- functional analysis: Hilbert spaces, linear operators, spectral theory
- differential equations: partial differential equations, separation of variables, ordinary differential equations, Sturm–Liouville theory, eigenfunctions
- harmonic analysis: Fourier transforms

HEISENBERG PICTURE

In physics, the Heisenberg picture (also called the Heisenberg representation) is a formulation (largely due to Werner Heisenberg in 1925) of quantum mechanics in which the operators (observables and others) incorporate a dependency on time, but the state vectors are time-independent, an arbitrary fixed basis rigidly underlying the theory.

It stands in contrast to the Schrödinger picture in which the operators are constant, instead, and the states evolve in time. The two pictures only differ by a basis change with respect to time-dependency, which corresponds to the difference between active and passive transformations. The Heisenberg picture is the formulation of matrix mechanics in an arbitrary basis, in which the Hamiltonian is not necessarily diagonal.

It further serves to define a third, hybrid, picture, the Interaction picture.

Mathematical Details

In the Heisenberg picture of quantum mechanics the state vectors, $|\psi(t)\rangle$, do not change with time, while observables A satisfy

$$\frac{d}{dt}A(t) = \frac{i}{\hbar}[H, A(t)] + \left(\frac{\partial A}{\partial t}\right)_H,$$

where H is the Hamiltonian and denotes the commutator of two operators (in this case H and A). Taking expectation values automatically yields the Ehrenfest theorem, featured in the correspondence principle.

By the Stone–von Neumann theorem, the Heisenberg picture and the Schrödinger picture are unitarily equivalent, just a basis change in Hilbert space. In some sense, the Heisenberg picture is more natural and convenient than the equivalent Schrödinger picture, especially for relativistic theories. Lorentz invariance is manifest in the Heisenberg picture, since the state vectors do not single out the time or space.

This approach also has a more direct similarity to classical physics: by simply replacing the commutator above by the Poisson bracket, the Heisenberg equation reduces to an equation in Hamiltonian mechanics.

Equivalence of Heisenberg's Equation to the Schrödinger Equation

For the sake of pedagogy, the Heisenberg picture is introduced here from the subsequent, but more familiar, Schrödinger picture. The expectation value of an observable A, which is a Hermitian linear operator, for a given Schrödinger state $|\psi(t)\rangle$, is given by

$$\langle A\rangle_t = \langle\psi(t)|A|\psi(t)\rangle.$$

In the Schrödinger picture, the state $|\psi(t)\rangle$ at time t is related to the state $|\psi(0)$ at time 0 by a unitary time-evolution operator, U(t),

$$|\psi(t)\rangle = U(t)|\psi(0)\rangle.$$

If the Hamiltonian does not vary with time, then the time-evolution operator can be written as

$$U(t) = e^{-iHt/\hbar},$$

where H is the Hamiltonian and ħ is the reduced Planck constant. Therefore,

$$\langle A\rangle_t = \langle\psi(0)|e^{iHt/\hbar}Ae^{-iHt/\hbar}|\psi(0)\rangle.$$

Peg all state vectors to a rigid basis of $|\psi(0)\rangle$ then, and define

$$A(t) := e^{iHt/\hbar}Ae^{-iHt/\hbar}.$$

It now follows that

$$\begin{aligned}\frac{d}{dt}A(t) &= \frac{i}{\hbar}He^{iHt/\hbar}Ae^{-iHt/\hbar} + e^{iHt/\hbar}\left(\frac{\partial A}{\partial t}\right)e^{-iHt/\hbar} + \frac{i}{\hbar}e^{iHt/\hbar}A\cdot(-H)e^{-iHt/\hbar} \\ &= \frac{i}{\hbar}e^{iHt/\hbar}\left(HA - AH\right)e^{-iHt/\hbar} + e^{iHt/\hbar}\left(\frac{\partial A}{\partial t}\right)e^{-iHt/\hbar} \\ &= \frac{i}{\hbar}\left(HA(t) - A(t)H\right) + e^{iHt/\hbar}\left(\frac{\partial A}{\partial t}\right)e^{-iHt/\hbar}.\end{aligned}$$

Differentiation was according to the product rule, while $\partial A/\partial t$ is the time derivative of the initial A, not the A(t) operator defined. The last equation holds since $\exp(-iHt/\hbar)$ commutes with H.

Thus

$$\frac{d}{dt}A(t) = \frac{i}{\hbar}[H, A(t)] + e^{iHt/\hbar}\left(\frac{\partial A}{\partial t}\right)e^{-iHt/\hbar},$$

and hence emerges the above Heisenberg equation of motion, since the convective functional dependence on x(0) and p(0) converts to the same dependence on x(t), p(t), so that the last term converts to $\partial A(t)/\partial t$. [X, Y] is the commutator of two operators and is defined as $[X, Y] := XY - YX$.

The equation is solved by the A(t) defined above, as evident by use of the standard operator identity,

$$e^BAe^{-B} = A + [B, A] + \frac{1}{2!}[B,[B,A]] + \frac{1}{3!}[B,[B,[B,A]]] + \cdots.$$

which implies

$$A(t) = A + \frac{it}{\hbar}[H, A] - \frac{t^2}{2!\hbar^2}[H,[H,A]] - \frac{it^3}{3!\hbar^3}[H,[H,[H,A]]] + \ldots$$

This relation also holds for classical mechanics, the classical limit of the above, given the correspondence between Poisson brackets and commutators,

$$[A, H] \leftrightarrow i\hbar\{A, H\}$$

In classical mechanics, for an A with no explicit time dependence,

$$\{A, H\} = \frac{d}{dt}A,$$

so, again, the expression for A(t) is the Taylor expansion around t = 0.

In effect, the arbitrary rigid Hilbert space basis $|\psi(0)\rangle$ has receded from view, and is only considered at the very last step of taking specific expectation values or matrix elements of observables.

Commutator Relations

Commutator relations may look different than in the Schrödinger picture, because of the time dependence of operators. For example, consider the operators $x(t_1)$, $x(t_2)$, $p(t_1)$ and $p(t_2)$. The time evolution of those operators depends on the Hamiltonian of the system. Considering the one-dimensional harmonic oscillator,

$$H = \frac{p^2}{2m} + \frac{m\omega^2 x^2}{2},$$

the evolution of the position and momentum operators is given by:

$$\frac{d}{dt}x(t) = \frac{i}{\hbar}[H, x(t)] = \frac{p}{m},$$

$$\frac{d}{dt}p(t) = \frac{i}{\hbar}[H, p(t)] = -m\omega^2 x.$$

Differentiating both equations once more and solving for them with proper initial conditions,

$$\dot{p}(0) = -m\omega^2 x_0,$$

$$\dot{x}(0) = \frac{p_0}{m},$$

leads to

$$x(t) = x_0 \cos(\omega t) + \frac{p_0}{\omega m}\sin(\omega t),$$

$$p(t) = p_0 \cos(\omega t) - m\omega x_0 \sin(\omega t).$$

Direct computation yields the more general commutator relations,

$$[x(t_1), x(t_2)] = \frac{i\hbar}{m\omega}\sin(\omega t_2 - \omega t_1),$$

$$[p(t_1), p(t_2)] = i\hbar m\omega \sin(\omega t_2 - \omega t_1),$$

$$[x(t_1), p(t_2)] = i\hbar \cos(\omega t_2 - \omega t_1).$$

For $t_1 = t_2$, one simply recovers the standard canonical commutation relations valid in all pictures.

INTERACTION PICTURE

In quantum mechanics, the **interaction picture** (also known as the **Dirac picture**) is an intermediate representation between the Schrödinger picture and the Heisenberg picture. Whereas in the other two pictures either the state vector or the operators carry time dependence, in the interaction picture both carry part of the time dependence of observables. The interaction picture is useful in dealing with changes to the wave functions and observable due to interactions. Most field theoretical calculations use the interaction representation because they construct the solution to the many body Schrödinger equation as the solution to the free particle problem plus some unknown interaction parts.

Equations that include operators acting at different times, which hold in the interaction picture, don't necessarily hold in the Schrödinger or the Heisenberg picture. This is because time-dependent unitary transformations relate operators in one picture to the analogous operators in the others.

Definition

Operators and state vectors in the interaction picture are related by a change of basis (unitary transformation) to those same operators and state vectors in the Schrödinger picture.

To switch into the interaction picture, we divide the Schrödinger picture Hamiltonian into two parts,

$$H_S = H_{0,S} + H_{1,S} \ .$$

Any possible choice of parts will yield a valid interaction picture; but in order for the interaction picture to be useful in simplifying the analysis of a problem, the parts will typically be chosen so that $H_{0,S}$ is well understood and exactly solvable, while $H_{1,S}$ contains some harder-to-analyze perturbation to this system.

If the Hamiltonian has explicit time-dependence (for example, if the quantum system interacts with an applied external electric field that varies in time), it will usually be advantageous to include the explicitly time-dependent terms with $H_{1,S}$, leaving $H_{0,S}$ time-independent. We proceed assuming that this is the case. If there is a context in which it makes sense to have $H_{0,S}$ be time-dependent, then one can proceed by replacing $e^{\pm iH_{0,S}t/\hbar}$ by the corresponding time-evolution operator in the definitions below.

State Vectors

A state vector in the interaction picture is defined as

$$|\psi_I(t)\rangle = e^{iH_{0,S}t/\hbar}|\psi_S(t)\rangle \ ,$$

where $|\psi_S(t)\rangle$ is the state vector in the Schrödinger picture.

Operators

An operator in the interaction picture is defined as

$$A_I(t) = e^{iH_{0,S}t/\hbar} A_S(t) e^{-iH_{0,S}t/\hbar} .$$

Note that $A_S(t)$ will typically not depend on t, and can be rewritten as just A_S. It only depends on t if the operator has "explicit time dependence", for example due to its dependence on an applied, external, time-varying electric field.

Hamiltonian Operator

For the operator H_0 itself, the interaction picture and Schrödinger picture coincide, $H_{0,I}(t) = e^{iH_{0,S}t/\hbar} H_{0,S} e^{-iH_{0,S}t/\hbar} = H_{0,S}$.

This is easily seen through the fact that operators commute with differentiable functions of themselves. This particular operator then can be called H_0 without ambiguity.

For the perturbation Hamiltonian $H_{1,I}$, however,

$$H_{1,I}(t) = e^{iH_{0,S}t/\hbar} H_{1,S} e^{-iH_{0,S}t/\hbar},$$

where the interaction picture perturbation Hamiltonian becomes a time-dependent Hamiltonian—unless $[H_{1,S}, H_{0,S}] = 0$.

It is possible to obtain the interaction picture for a time-dependent Hamiltonian $H_{0,S}(t)$ as well, but the exponentials need to be replaced by the unitary propagator for the evolution generated by $H_{0,S}(t)$, or more explicitly with a time-ordered exponential integral.

Density Matrix

The density matrix can be shown to transform to the interaction picture in the same way as any other operator. In particular, let ρ_I and ρ_S be the density matrix in the interaction picture and the Schrödinger picture, respectively. If there is probability p_n to be in the physical state $|\psi_n\rangle$, then

$$\rho_I(t) = \sum_n p_n(t) |\psi_{n,I}(t)\rangle\langle\psi_{n,I}(t)| = \sum_n p_n(t) e^{iH_{0,S}t/\hbar} |\psi_{n,S}(t)\rangle\langle\psi_{n,S}(t)| e^{-iH_{0,S}t/\hbar} = e^{iH_{0,S}t/\hbar} \rho_S(t) e^{-iH_{0,S}t/\hbar}.$$

Evolution	Picture		
of:	Heisenberg	Interaction	Schrödinger
Ket state	constant	$\vert\psi_I(t)\rangle = e^{iH_{0,S}\,t/\hbar}\vert\psi_S(t)\rangle$	$\vert\psi_S(t)\rangle = e^{-iH_S\,t/\hbar}\vert\psi_S(0)\rangle$
Observable	$A_H(t) = e^{iH_S\,t/\hbar} A_S e^{-iH_S\,t/\hbar}$	$A_I(t) = e^{iH_{0,S}\,t/\hbar} A_S e^{-iH_{0,S}\,t/\hbar}$	constant
Density matrix	constant	$\rho_I(t) = e^{iH_{0,S}\,t/\hbar}\rho_S(t)e^{-iH_{0,S}\,t/\hbar}$	$\rho_S(t) = e^{-iH_S\,t/\hbar}\rho_S(0)e^{iH_S\,t/\hbar}$

Time-Evolution Equations in the Interaction Picture

Time-Evolution Of States

Transforming the Schrödinger equation into the interaction picture gives:

$$i\hbar\frac{d}{dt} \mid \psi_I(t)\rangle = H_{1,I}(t) \mid \psi_I(t)\rangle.$$

This equation is referred to as the **Schwinger–Tomonaga equation**.

Time-Evolution of Operators

If the operator A_S is time independent (i.e., does not have "explicit time dependence"; see above), then the corresponding time evolution for $A_I(t)$ is given by

$$i\hbar\frac{d}{dt}A_I(t) = [A_I(t), H_{0,S}] .$$

In the interaction picture the operators evolve in time like the operators in the Heisenberg picture with the Hamiltonian H' $=H_0$.

Time-Evolution of the Density Matrix

Transforming the Schwinger–Tomonaga equation into the language of the density matrix (or equivalently, transforming the von Neumann equation into the interaction picture) gives:

$$i\hbar\frac{d}{dt}\rho_I(t) = [H_{1,I}(t), \rho_I(t)] .$$

Expectation Values

For a general operator A, the expectation value in the interaction picture is given by

$$\langle A_I(t)\rangle = \langle\psi_I(t)|A_I(t)|\psi_I(t)\rangle = \langle\psi_S(t)|e^{-iH_{0,S}t}e^{iH_{0,S}t}\,A_S\,e^{-iH_{0,S}t}e^{iH_{0,S}t}|\psi_S(t)\rangle = \langle A_S(t)\rangle$$

Using the density matrix expression for expectation value, we will get

$$\langle A_I(t)\rangle = Tr(\rho_I(t)\,A_I(t))$$

Use of Interaction Picture

The purpose of the interaction picture is to shunt all the time dependence due to H_0 onto the operators, thus allowing them to evolve freely, and leaving only $H_{1,I}$ to control the time-evolution of the state vectors.

The interaction picture is convenient when considering the effect of a small interaction term, $H_{1,S}$, being added to the Hamiltonian of a solved system, $H_{0,S}$. By utilizing the interaction picture, one can use time-dependent perturbation theory to find the effect of $H_{1,I}$, e.g., in the derivation of Fermi's golden rule, or the Dyson series, in quantum field theory: In 1947, Tomonaga and Schwinger appreciated that covariant perturbation theory could be formulated elegantly in the interaction picture, since field operators can evolve in time as free fields, even in the presence of interactions, now treated perturbatively in such a Dyson series.

MATRIX MECHANICS

Matrix mechanics is a formulation of quantum mechanics created by Werner Heisenberg, Max Born, and Pascual Jordan in 1925.

Matrix mechanics was the first conceptually autonomous and logically consistent formulation of quantum mechanics. Its account of quantum jumps supplanted the Bohr Model's electron orbits. It did so by interpreting the physical properties of particles as matrices that evolve in time. It is equivalent to the Schrödinger wave formulation of quantum mechanics, as manifest in Dirac's bra–ket notation.

In some contrast to the wave formulation, it produces spectra of energy operators by purely algebraic, ladder operator, methods. Relying on these methods, Pauli derived the hydrogen atom spectrum in 1926, before the development of wave mechanics.

Development of Matrix Mechanics

In 1925, Werner Heisenberg, Max Born, and Pascual Jordan formulated the matrix mechanics representation of quantum mechanics.

Epiphany at Helgoland

In 1925 Werner Heisenberg was working in Göttingen on the problem of calculating the spectral lines of hydrogen. By May 1925 he began trying to describe atomic systems by observables only. On June 7, to escape the effects of a bad attack of hay fever, Heisenberg left for the pollen free North Sea island of Helgoland. While there, in between climbing and learning by heart poems from Goethe's West-östlicher Diwan, he continued to ponder the spectral issue and eventually realised that adopting non-commuting observables might solve the problem, and he later wrote

It was about three o' clock at night when the final result of the calculation lay before me. At first I was deeply shaken. I was so excited that I could not think of sleep. So I left the house and awaited the sunrise on the top of a rock.

The Three Fundamental Papers

After Heisenberg returned to Göttingen, he showed Wolfgang Pauli his calculations, commenting at one point:

"Everything is still vague and unclear to me, but it seems as if the electrons will no more move on orbits."

On July 9 Heisenberg gave the same paper of his calculations to Max Born, saying, "...he had written a crazy paper and did not dare to send it in for publication, and that Born should read it and advise him on it..." prior to publication. Heisenberg then departed for a while, leaving Born to analyse the paper.

In the paper, Heisenberg formulated quantum theory without sharp electron orbits. Hendrik Kramers had earlier calculated the relative intensities of spectral lines in the Sommerfeld model by interpreting the Fourier coefficients of the orbits as intensities. But his answer, like all other calculations in the old quantum theory, was only correct for large orbits.

Heisenberg, after a collaboration with Kramers, began to understand that the transition probabilities were not quite classical quantities, because the only frequencies that appear in the Fourier series should be the ones that are observed in quantum jumps, not the fictional ones that come from Fourier-analyzing sharp classical orbits. He replaced the classical Fourier series with a matrix of coefficients, a fuzzed-out quantum analog of the Fourier series. Classically, the Fourier coefficients give the intensity of the emitted radiation, so in quantum mechanics the magnitude of the matrix elements of the position operator were the intensity of radiation in the bright-line spectrum. The quantities in Heisenberg's formulation were the classical position and momentum, but now they were no longer sharply defined. Each quantity was represented by a collection of Fourier coefficients with two indices, corresponding to the initial and final states.

When Born read the paper, he recognized the formulation as one which could be transcribed and extended to the systematic language of matrices, which he had learned from his study under Jakob Rosanes at Breslau University. Born, with the help of his assistant and former student Pascual Jordan, began immediately to make the transcription and extension, and they submitted their results for publication; the paper was received for publication just 60 days after Heisenberg's paper.

A follow-on paper was submitted for publication before the end of the year by all three authors. (A brief review of Born's role in the development of the matrix mechanics formulation of quantum mechanics along with a discussion of the key formula involving the non-commutivity of the probability amplitudes can be found in an article by Jeremy Bernstein. A detailed historical and technical account can be found in Mehra and Rechenberg's book The Historical Development of Quantum Theory. Volume 3. The Formulation of Matrix Mechanics and Its Modifications 1925–1926.)

Up until this time, matrices were seldom used by physicists; they were considered to belong to the realm of pure mathematics. Gustav Mie had used them in a paper on electrodynamics in 1912 and Born had used them in his work on the lattices theory of crystals in 1921. While matrices were used in these cases, the algebra of matrices with their multiplication did not enter the picture as they did in the matrix formulation of quantum mechanics.

Born, however, had learned matrix algebra from Rosanes, as already noted, but Born had also learned Hilbert's theory of integral equations and quadratic forms for an infinite number of variables as was apparent from a citation by Born of Hilbert's work Grundzüge einer allgemeinen Theorie der Linearen Integralgleichungen published in 1912.

Jordan, too was well equipped for the task. For a number of years, he had been an assistant to Richard Courant at Göttingen in the preparation of Courant and David Hilbert's book Methoden der mathematischen Physik I, which was published in 1924. This book, fortuitously, contained a great many of the mathematical tools necessary for the continued development of quantum mechanics.

In 1926, John von Neumann became assistant to David Hilbert, and he would coin the term Hilbert space to describe the algebra and analysis which were used in the development of quantum mechanics.

Heisenberg's Reasoning

Before matrix mechanics, the old quantum theory described the motion of a particle by a classical orbit, with well defined position and momentum X(t), P(t), with the restriction that the time integral over one period T of the momentum times the velocity must be a positive integer multiple of Planck's constant

$$\int_0^T P\,dX = nh\,.$$

While this restriction correctly selects orbits with more or less the right energy values E_n, the old quantum mechanical formalism did not describe time dependent processes, such as the emission or absorption of radiation.

When a classical particle is weakly coupled to a radiation field, so that the radiative damping can be neglected, it will emit radiation in a pattern which repeats itself every orbital period. The frequencies which make up the outgoing wave are then integer multiples of the orbital frequency, and this is a reflection of the fact that X(t) is periodic, so that its Fourier representation has frequencies $2\pi n/T$ only.

$$X(t) = \sum_{n=-\infty}^{\infty} e^{2\pi int/T} X_n\,.$$

The coefficients X_n are complex numbers. The ones with negative frequencies must be the complex conjugates of the ones with positive frequencies, so that X(t) will always be real,

$$X_n = X^*_{-n}\,.$$

A quantum mechanical particle, on the other hand, can't emit radiation continuously, it can only emit photons. Assuming that the quantum particle started in orbit number n, emitted a photon, then ended up in orbit number m, the energy of the photon is $E_n - E_m$, which means that its frequency is $(E_n - E_m)/h$.

For large n and m, but with n−m relatively small, these are the classical frequencies by Bohr's correspondence principle

$$E_n - E_m \approx h(n-m)/T\,.$$

In the formula above, T is the classical period of either orbit n or orbit m, since the difference between them is higher order in h. But for n and m small, or if n − m is large, the frequencies are not integer multiples of any single frequency.

Since the frequencies which the particle emits are the same as the frequencies in the fourier description of its motion, this suggests that something in the time-dependent description of the particle is oscillating with frequency $(E_n - E_m)/h$. Heisenberg called this quantity X_{nm}, and demanded that it should reduce to the classical Fourier coefficients in the classical limit. For large values of n, m but with n − m relatively small, X_{nm} is the (n−m)th Fourier coefficient of the classical motion at orbit n. Since X_{nm} has opposite frequency to X_{mn}, the condition that X is real becomes

$$X_{nm} = X^*_{mn} .$$

By definition, X_{nm} only has the frequency $(E_n - E_m)/h$, so its time evolution is simple:

$$X_{nm}(t) = e^{2\pi i(E_n - E_m)t/h} X_{nm}(0) .$$

This is the original form of Heisenberg's equation of motion.

Given two arrays X_{nm} and P_{nm} describing two physical quantities, Heisenberg could form a new array of the same type by combining the terms $X_{nk}P_{km}$, which also oscillate with the right frequency. Since the Fourier coefficients of the product of two quantities is the convolution of the Fourier coefficients of each one separately, the correspondence with Fourier series allowed Heisenberg to deduce the rule by which the arrays should be multiplied,

$$(XP)_{mn} = \sum_{k=0}^{\infty} X_{mk} P_{kn} .$$

Born pointed out that this is the law of matrix multiplication, so that the position, the momentum, the energy, all the observable quantities in the theory, are interpreted as matrices. Under this multiplication rule, the product depends on the order: XP is different from PX.

The X matrix is a complete description of the motion of a quantum mechanical particle. Because the frequencies in the quantum motion are not multiples of a common frequency, the matrix elements cannot be interpreted as the Fourier coefficients of a sharp classical trajectory. Nevertheless, as matrices, X(t) and P(t) satisfy the classical equations of motion; also see Ehrenfest's theorem, below.

Matrix Basics

When it was introduced by Werner Heisenberg, Max Born and Pascual Jordan in 1925, matrix mechanics was not immediately accepted and was a source of controversy, at first. Schrödinger's later introduction of wave mechanics was greatly favored.

Part of the reason was that Heisenberg's formulation was in an odd mathematical language, for the time, while Schrödinger's formulation was based on familiar wave equations. But there was also a deeper sociological reason. Quantum mechanics had been developing by two paths, one under the direction of Einstein and the other under the direction of Bohr. Einstein emphasized wave–particle duality, while Bohr emphasized the discrete energy states and quantum jumps. De Broglie had shown how to reproduce the discrete energy states in Einstein's framework – the quantum condition is the standing wave condition, and this gave hope to those in the Einstein school that all the discrete aspects of quantum mechanics would be subsumed into a continuous wave mechanics.

Matrix mechanics, on the other hand, came from the Bohr school, which was concerned with discrete energy states and quantum jumps. Bohr's followers did not appreciate physical models which pictured electrons as waves, or as anything at all. They preferred to focus on the quantities which were directly connected to experiments.

In atomic physics, spectroscopy gave observational data on atomic transitions arising from the interactions of atoms with light quanta. The Bohr school required that only those quantities which were in principle measurable by spectroscopy should appear in the theory. These quantities include the energy levels and their intensities but they do not include the exact location of a particle in its Bohr orbit. It is very hard to imagine an experiment which could determine whether an electron in the ground state of a hydrogen atom is to the right or to the left of the nucleus. It was a deep conviction that such questions did not have an answer.

The matrix formulation was built on the premise that all physical observables are represented by matrices, whose elements are indexed by two different energy levels. The set of eigenvalues of the matrix were eventually understood to be the set of all possible values that the observable can have. Since Heisenberg's matrices are Hermitian, the eigenvalues are real.

If an observable is measured and the result is a certain eigenvalue, the corresponding eigenvector is the state of the system immediately after the measurement. The act of measurement in matrix mechanics 'collapses' the state of the system. If one measures two observables simultaneously, the state of the system collapses to a common eigenvector of the two observables. Since most matrices don't have any eigenvectors in common, most observables can never be measured precisely at the same time. This is the uncertainty principle.

If two matrices share their eigenvectors, they can be simultaneously diagonalized. In the basis where they are both diagonal, it is clear that their product does not depend on their order because multiplication of diagonal matrices is just multiplication of numbers. The uncertainty principle, by contrast, is an expression of the fact that often two matrices A and B do not always commute, i.e., that AB – BA does not necessarily equal 0. The fundamental commutation relation of matrix mechanics,

$$\sum_k (X_{nk}P_{km} - P_{nk}X_{km}) = \frac{ih}{2\pi}\,\delta_{nm}$$

mplies then that there are no states which simultaneously have a definite position and momentum.

This principle of uncertainty holds for many other pairs of observables as well. For example, the energy does not commute with the position either, so it is impossible to precisely determine the position and energy of an electron in an atom.

Nobel Prize

In 1928, Albert Einstein nominated Heisenberg, Born, and Jordan for the Nobel Prize in Physics. The announcement of the Nobel Prize in Physics for 1932 was delayed until November 1933. It was at that time that it was announced Heisenberg had won the Prize for 1932 "for the creation of quantum mechanics, the application of which has, inter alia, led to the discovery of the allotropic forms of hydrogen" and Erwin Schrödinger and Paul Adrien Maurice Dirac shared the 1933 Prize "for the discovery of new productive forms of atomic theory".

One can rightly ask why Born was not awarded the Prize in 1932 along with Heisenberg, and Bernstein gives some speculations on this matter. One of them is related to Jordan joining the Nazi Party on May 1, 1933 and becoming a Storm Trooper. Hence, Jordan's Party affiliations and Jordan's links to Born may have affected Born's chance at the Prize at that time. Bernstein also notes that when Born won the Prize in 1954, Jordan was still alive, and the Prize was awarded for the statistical interpretation of quantum mechanics, attributable to Born alone.

Heisenberg's reactions to Born for Heisenberg receiving the Prize for 1932 and for Born receiving the Prize in 1954 are also instructive in evaluating whether Born should have shared the Prize with Heisenberg. On November 25, 1933 Born received a letter from Heisenberg in which he said he had been delayed in writing due to a "bad conscience" that he alone had received the Prize "for work done in Göttingen in collaboration – you, Jordan and I." Heisenberg went on to say that Born and Jordan's contribution to quantum mechanics cannot be changed by "a wrong decision from the outside."

In 1954, Heisenberg wrote an article honoring Max Planck for his insight in 1900. In the article, Heisenberg credited Born and Jordan for the final mathematical formulation of matrix mechanics and Heisenberg went on to stress how great their contributions were to quantum mechanics, which were not "adequately acknowledged in the public eye."

Mathematical Development

Once Heisenberg introduced the matrices for X and P, he could find their matrix elements in special cases by guesswork, guided by the correspondence principle. Since the matrix elements are the quantum mechanical analogs of Fourier coefficients of the classical orbits, the simplest case is the harmonic oscillator, where the classical position and momentum, X(t) and P(t), are sinusoidal.

Harmonic Oscillator

In units where the mass and frequency of the oscillator are equal to one (see nondimensionalization), the energy of the oscillator is

$$H = \frac{1}{2}(P^2 + X^2)\,.$$

The level sets of H are the clockwise orbits, and they are nested circles in phase space. The classical orbit with energy E is

$$X(t) = \sqrt{2E}\cos(t), \qquad P(t) = -\sqrt{2E}\sin(t)\,.$$

The old quantum condition dictates that the integral of P dX over an orbit, which is the area of the circle in phase space, must be an integer multiple of Planck's constant. The area of the circle of radius $\sqrt{2E}$ is $2\pi E$. So

$$E = \frac{nh}{2\pi}\,,$$

or, in natural units where $\hbar = 1$, the energy is an integer.

The Fourier components of X(t) and P(t) are simple, and more so if they are combined into the quantities

$$A(t) = X(t) + iP(t) = \sqrt{2E}\,e^{-it}, \quad A^\dagger(t) = X(t) - iP(t) = \sqrt{2E}\,e^{it}.$$

Both A and $A^\dagger$ have only a single frequency, and X and P can be recovered from their sum and difference.

Since A(t) has a classical Fourier series with only the lowest frequency, and the matrix element A_{mn} is the (m − n)th Fourier coefficient of the classical orbit, the matrix for A is nonzero only on the line just above the diagonal, where it is equal to $\sqrt{2E_n}$. The matrix for $A^\dagger$ is likewise only nonzero on the line below the diagonal, with the same elements.

Thus, from A and $A^\dagger$, reconstruction yields

$$\sqrt{2}X(0) = \sqrt{\frac{h}{2\pi}} \begin{bmatrix} 0 & \sqrt{1} & 0 & 0 & 0 & \cdots \\ \sqrt{1} & 0 & \sqrt{2} & 0 & 0 & \cdots \\ 0 & \sqrt{2} & 0 & \sqrt{3} & 0 & \cdots \\ 0 & 0 & \sqrt{3} & 0 & \sqrt{4} & \cdots \\ \vdots & \vdots & \vdots & \vdots & \vdots & \ddots \end{bmatrix},$$

and

$$\sqrt{2}P(0) = \sqrt{\frac{h}{2\pi}} \begin{bmatrix} 0 & -i\sqrt{1} & 0 & 0 & 0 & \cdots \\ i\sqrt{1} & 0 & -i\sqrt{2} & 0 & 0 & \cdots \\ 0 & i\sqrt{2} & 0 & -i\sqrt{3} & 0 & \cdots \\ 0 & 0 & i\sqrt{3} & 0 & -i\sqrt{4} & \cdots \\ \vdots & \vdots & \vdots & \vdots & \vdots & \ddots \end{bmatrix},$$

which, up to the choice of units, are the Heisenberg matrices for the harmonic oscillator. Note that both matrices are hermitian, since they are constructed from the Fourier coefficients of real quantities.

Finding X(t) and P(t) is direct, since they are quantum Fourier coefficients so they evolve simply with time,

$$X_{mn}(t) = X_{mn}(0)e^{i(E_m - E_n)t}, \quad P_{mn}(t) = P_{mn}(0)e^{i(E_m - E_n)t}.$$

The matrix product of X and P is not hermitian, but has a real and imaginary part. The real part is one half the symmetric expression XP + PX, while the imaginary part is proportional to the commutator

$$[X, P] = (XP - PX).$$

It is simple to verify explicitly that XP − PX in the case of the harmonic oscillator, is iħ, multiplied by the identity.

It is likewise simple to verify that the matrix

$$H = \frac{1}{2}(X^2 + P^2)$$

is a diagonal matrix, with eigenvalues E_i.

Conservation of Energy

The harmonic oscillator is an important case. Finding the matrices is easier than determining the general conditions from these special forms. For this reason, Heisenberg investigated the anharmonic oscillator, with Hamiltonian

$$H = \frac{1}{2}P^2 + \frac{1}{2}X^2 + \epsilon X^3 .$$

In this case, the X and P matrices are no longer simple off diagonal matrices, since the corresponding classical orbits are slightly squashed and displaced, so that they have Fourier coefficients at every classical frequency. To determine the matrix elements, Heisenberg required that the classical equations of motion be obeyed as matrix equations,

$$\frac{dX}{dt} = P \quad \frac{dP}{dt} = -X - 3\epsilon X^2 .$$

He noticed that if this could be done, then H, considered as a matrix function of X and P, will have zero time derivative.

$$\frac{dH}{dt} = P * \frac{dP}{dt} + (X + 3\epsilon X^2) * \frac{dX}{dt} = 0 ,$$

where A*B is the anticommutator,

$$A * B = \frac{1}{2}(AB + BA) .$$

Given that all the off diagonal elements have a nonzero frequency; H being constant implies that H is diagonal. It was clear to Heisenberg that in this system, the energy could be exactly conserved in an arbitrary quantum system, a very encouraging sign.

The process of emission and absorption of photons seemed to demand that the conservation of energy will hold at best on average. If a wave containing exactly one photon passes over some atoms, and one of them absorbs it, that atom needs to tell the others that they can't absorb the photon anymore. But if the atoms are far apart, any signal cannot reach the other atoms in time, and they might end up absorbing the same photon anyway and dissipating the energy to the environment. When the signal reached them, the other atoms would have to somehow recall that energy. This paradox led Bohr, Kramers and Slater to abandon exact conservation of energy. Heisenberg's formalism, when extended to include the electromagnetic field, was obviously going to sidestep this problem, a hint that the interpretation of the theory will involve wavefunction collapse.

Differentiation Trick — Canonical Commutation Relations

Demanding that the classical equations of motion are preserved is not a strong enough condition to determine the matrix elements. Planck's constant does not appear in the classical equations, so that the matrices could be constructed for many different values of ħ and still satisfy the equations of motion, but with different energy levels.

So, in order to implement his program, Heisenberg needed to use the old quantum condition to fix the energy levels, then fill in the matrices with Fourier coefficients of the classical equations, then alter the matrix coefficients and the energy levels slightly to make sure the classical equations are satisfied. This is clearly not satisfactory. The old quantum conditions refer to the area enclosed by the sharp classical orbits, which do not exist in the new formalism.

The most important thing that Heisenberg discovered is how to translate the old quantum condition into a simple statement in matrix mechanics.

To do this, he investigated the action integral as a matrix quantity,

$$\int_0^T \sum_k P_{mk}(t)\frac{dX_{kn}}{dt}dt \overset{?}{\approx} J_{mn} \,.$$

There are several problems with this integral, all stemming from the incompatibility of the matrix formalism with the old picture of orbits. Which period T should be used? Semiclassically, it should be either m or n, but the difference is order ħ, and an answer to order ħ is sought. The quantum condition tells us that J_{mn} is $2\pi n$ on the diagonal, so the fact that J is classically constant tells us that the off-diagonal elements are zero.

His crucial insight was to differentiate the quantum condition with respect to n. This idea only makes complete sense in the classical limit, where n is not an integer but the continuous action variable J, but Heisenberg performed analogous manipulations with matrices, where the intermediate expressions are sometimes discrete differences and sometimes derivatives.

In the following discussion, for the sake of clarity, the differentiation will be performed on the classical variables, and the transition to matrix mechanics will be done afterwards, guided by the correspondence principle.

In the classical setting, the derivative is the derivative with respect to J of the integral which defines J, so it is tautologically equal to 1.

$$\frac{d}{dJ}\int_0^T PdX = 1$$

$$= \int_0^T dt\left(\frac{dP}{dJ}\frac{dX}{dt} + P\frac{d}{dJ}\frac{dX}{dt}\right) = \int_0^T dt\left(\frac{dP}{dJ}\frac{dX}{dt} - \frac{dP}{dt}\frac{dX}{dJ}\right)$$

where the derivatives dP/dJ and dX/dJ should be interpreted as differences with respect to J at corresponding times on nearby orbits, exactly what would be obtained if the Fourier coefficients of the orbital motion were differentiated. (These derivatives are symplectically orthogonal in phase space to the time derivatives dP/dt and dX/dt).

The final expression is clarified by introducing the variable canonically conjugate to J, which is called the angle variable θ: The derivative with respect to time is a derivative with respect to θ, up to a factor of 2πT,

$$\frac{2\pi}{T}\int_0^T dt\left(\frac{dp}{dJ}\frac{dX}{d\theta} - \frac{dP}{d\theta}\frac{dX}{dJ}\right) = 1\,.$$

So the quantum condition integral is the average value over one cycle of the Poisson bracket of X and P.

An analogous differentiation of the Fourier series of P dX demonstrates that the off-diagonal elements of the Poisson bracket are all zero. The Poisson bracket of two canonically conjugate variables, such as X and P, is the constant value 1, so this integral really is the average value of 1; so it is 1, as we knew all along, because it is dJ/dJ after all. But Heisenberg, Born and Jordan, unlike Dirac, were not familiar with the theory of Poisson brackets, so, for them, the differentiation effectively evaluated {X, P} in J, θ coordinates.

The Poisson Bracket, unlike the action integral, does have a simple translation to matrix mechanics—–it normally corresponds to the imaginary part of the product of two variables, the commutator.

To see this, examine the (antisymmetrized) product of two matrices A and B in the correspondence limit, where the matrix elements are slowly varying functions of the index, keeping in mind that the answer is zero classically.

In the correspondence limit, when indices m, n are large and nearby, while k,r are small, the rate of change of the matrix elements in the diagonal direction is the matrix element of the J derivative of the corresponding classical quantity. So its possible to shift any matrix element diagonally through the correspondence,

$$A_{(m+r)(n+r)} - A_{mn} \approx r \left(\frac{dA}{dJ}\right)_{mn}$$

where the right hand side is really only the (m − n)'th Fourier component of dA/dJ at the orbit near m to this semiclassical order, not a full well-defined matrix.

The semiclassical time derivative of a matrix element is obtained up to a factor of i by multiplying by the distance from the diagonal,

$$ikA_{m(m+k)} \approx \left(\frac{T}{2\pi}\frac{dA}{dt}\right)_{m(m+k)} = \left(\frac{dA}{d\theta}\right)_{m(m+k)}.$$

since the coefficient $A_{m(m+k)}$ is semiclassically the k'th Fourier coefficient of the m-th classical orbit.

The imaginary part of the product of A and B can be evaluated by shifting the matrix elements around so as to reproduce the classical answer, which is zero.

The leading nonzero residual is then given entirely by the shifting. Since all the matrix elements are at indices which have a small distance from the large index position (m,m), it helps to introduce two temporary notations: $A[r,k] = A_{(m+r)(m+k)}$ for the matrices, and (dA/dJ)[r] for the r'th Fourier components of classical quantities,

$$(AB-BA)[0,k]=\sum_{r=-\infty}^{\infty}(A[0,r]B[r,k]-A[r,k]B[0,r])$$

$$=\sum_{r}\left(A[-r+k,k]+(r-k)\frac{dA}{dJ}[r]\right)\left(B[0,k-r]+r\frac{dB}{dJ}[r-k]\right)-\sum_{r}A[r,k]B[0,r].$$

Flipping the summation variable in the first sum from r to $r' = k - r$, the matrix element becomes,

$$\sum_{r'}\left(A[r',k]-r'\frac{dA}{dJ}[k-r']\right)\left(B[0,r']+(k-r')\frac{dB}{dJ}[r']\right)-\sum_{r}A[r,k]B[0,r]$$

and it is clear that the principal (classical) part cancels.

The leading quantum part, neglecting the higher order product of derivatives in the residual expression, is then

$$=\sum_{r'}\left(\frac{dB}{dJ}[r'](k-r')A[r',k]-\frac{dA}{dJ}[k-r']r'B[0,r']\right)$$

so that, finally,

$$(AB-BA)[0,k]=\sum_{r'}\left(\frac{dB}{dJ}[r']i\frac{dA}{d\theta}[k-r']-\frac{dA}{dJ}[k-r']i\frac{dB}{d\theta}[r']\right)$$

which can be identified with i times the k-th classical Fourier component of the Poisson bracket.

Heisenberg's original differentiation trick was eventually extended to a full semiclassical derivation of the quantum condition, in collaboration with Born and Jordan. Once they were able to establish that

$$\frac{ih}{2\pi}\{X,P\}_{\mathrm{PB}} \qquad \longmapsto \qquad [X,P]\equiv XP-PX=\frac{ih}{2\pi},$$

this condition replaced and extended the old quantization rule, allowing the matrix elements of P and X for an arbitrary system to be determined simply from the form of the Hamiltonian.

The new quantization rule was assumed to be universally true, even though the derivation from the old quantum theory required semiclassical reasoning. (A full quantum treatment, however, for more elaborate arguments of the brackets, was appreciated in the 1940s to amount to extending Poisson brackets to Moyal brackets.)

State Vectors and the Heisenberg Equation

To make the transition to standard quantum mechanics, the most important further addition was the quantum state vector, now written $|\psi\rangle$, which is the vector that the matrices act on. Without the state vector, it is not clear which particular motion the Heisenberg matrices are describing, since they include all the motions somewhere.

The interpretation of the state vector, whose components are written ψ_m, was furnished by Born. This interpretation is statistical: the result of a measurement of the physical quantity corresponding to the matrix A is random, with an average value equal to

$$\sum_{mn} \psi_m^* A_{mn} \psi_n \,.$$

Alternatively, and equivalently, the state vector gives the probability amplitude ψ_n for the quantum system to be in the energy state n.

Once the state vector was introduced, matrix mechanics could be rotated to any basis, where the H matrix need no longer be diagonal. The Heisenberg equation of motion in its original form states that A_{mn} evolves in time like a Fourier component,

$$A_{mn}(t) = e^{i(E_m - E_n)t} A_{mn}(0) \,,$$

which can be recast in differential form

$$\frac{dA_{mn}}{dt} = i(E_m - E_n) A_{mn} \,,$$

and it can be restated so that it is true in an arbitrary basis, by noting that the H matrix is diagonal with diagonal values E_m,

$$\frac{dA}{dt} = i(HA - AH) \,.$$

This is now a matrix equation, so it holds in any basis. This is the modern form of the Heisenberg equation of motion.

Its formal solution is:

$$A(t) = e^{iHt} A(0) e^{-iHt} \,.$$

All these forms of the equation of motion above say the same thing, that A(t) is equivalent to A(0), through a basis rotation by the unitary matrix e^{Ht}, a systematic picture elucidated by Dirac in his bra–ket notation.

Conversely, by rotating the basis for the state vector at each time by e^{iHt}, the time dependence in the matrices can be undone. The matrices are now time independent, but the state vector rotates,

$$|\psi(t)\rangle = e^{-iHt} |\psi(0)\rangle, \qquad \frac{d|\psi\rangle}{dt} = -iH|\psi\rangle \,.$$

This is the Schrödinger equation for the state vector, and this time-dependent change of basis amounts to transformation to the Schrödinger picture, with $\langle x|\psi\rangle = \psi(x)$.

In quantum mechanics in the Heisenberg picture the state vector, $|\psi\rangle$ does not change with time, while an observable A satisfies the ***Heisenberg equation of motion***,

$$\frac{dA}{dt} = \frac{i}{\hbar}[H, A] + \frac{\partial A}{\partial t} \,.$$

The extra term is for operators such as

$$A = (X + t^2 P)$$

which have an explicit time dependence, in addition to the time dependence from the unitary evolution discussed.

The Heisenberg picture does not distinguish time from space, so it is better suited to relativistic theories than the Schrödinger equation. Moreover, the similarity to classical physics is more manifest: the Hamiltonian equations of motion for classical mechanics are recovered by replacing the commutator above by the Poisson bracket . By the Stone–von Neumann theorem, the Heisenberg picture and the Schrödinger picture must be unitarily equivalent, as detailed below.

Further Results

Matrix mechanics rapidly developed into modern quantum mechanics, and gave interesting physical results on the spectra of atoms.

Wave Mechanics

Jordan noted that the commutation relations ensure that P acts as a differential operator. The operator identity

$$[a, bc] = abc - bca = abc - bac + bac - bca = [a, b]c + b[a, c]$$

allows the evaluation of the commutator of P with any power of X, and it implies that

$$[P, X^n] = -in\, X^{n-1}$$

which, together with linearity, implies that a P-commutator effectively differentiates any analytic matrix function of X.

Assuming limits are defined sensibly, this extends to arbitrary functions--but the extension need not be made explicit until a certain degree of mathematical rigor is required,

$$[P, f(X)] = -if'(X) .$$

Since X is a Hermitian matrix, it should be diagonalizable, and it will be clear from the eventual form of P that every real number can be an eigenvalue. This makes some of the mathematics subtle, since there is a separate eigenvector for every point in space.

In the basis where X is diagonal, an arbitrary state can be written as a superposition of states with eigenvalues x,

$$|\psi\rangle = \int_x \psi(x)|x\rangle \ ,$$

so that $\psi(x) = \langle x|\psi\rangle$, and the operator X multiplies each eigenvector by x,

$$X|\psi\rangle = \int_x x\psi(x)|x\rangle \ .$$

Define a linear operator D which differentiates ψ,

$$D\int_x \psi(x)|x\rangle = \int_x \psi'(x)|x\rangle \ ,$$

and note that

$$(DX - XD)|\psi\rangle = \int_x \left[(x\psi(x))' - x\psi'(x)\right]|x\rangle = \int_x \psi(x)|x\rangle = |\psi\rangle \ ,$$

so that the operator $-iD$ obeys the same commutation relation as P. Thus, the difference between P and $-iD$ must commute with X,

$$[P + iD, X] = 0 \ ,$$

so it may be simultaneously diagonalized with X: its value acting on any eigenstate of X is some function f of the eigenvalue x.

This function must be real, because both P and −iD are Hermitian,

$$(P + iD)|x\rangle = f(x)|x\rangle \ ,$$

rotating each state $|x\rangle$ by a phase $f(x)$, that is, redefining the phase of the wavefunction:

$$\psi(x) \to e^{-if(x)}\psi(x) \ .$$

The operator iD is redefined by an amount:

$$iD \to iD + f(X) \ ,$$

which means that, in the rotated basis, P is equal to $-iD$.

Hence, there is always a basis for the eigenvalues of X where the action of P on any wavefunction is known:

$$P\int_x \psi(x)|x\rangle = \int_x -i\psi'(x)|x\rangle \ ,$$

and the Hamiltonian in this basis is a linear differential operator on the state-vector components,

$$\left[\frac{P^2}{2m} + V(X)\right]\int_x \psi_x|x\rangle = \int_x \left[-\frac{1}{2m}\frac{\partial^2}{\partial x^2} + V(x)\right]\psi_x|x\rangle$$

Thus, the equation of motion for the state vector is but a celebrated differential equation,

$$i\frac{\partial}{\partial t}\psi_t(x) = \left[-\frac{1}{2m}\frac{\partial^2}{\partial x^2} + V(x)\right]\psi_t(x) \ .$$

Since D is a differential operator, in order for it to be sensibly defined, there must be eigenvalues of X which neighbors every given value. This suggests that the only possibility is that the space of all eigenvalues of X is all real numbers, and that P is iD, up to a phase rotation.

To make this rigorous requires a sensible discussion of the limiting space of functions, and in this space this is the Stone–von Neumann theorem: any operators X and P which obey the commutation relations can be made to act on a space of wavefunctions, with P a derivative operator. This implies that a Schrödinger picture is always available.

Matrix mechanics easily extends to many degrees of freedom in a natural way. Each degree of freedom has a separate X operator and a separate effective differential operator P, and the wavefunction is a function of all the possible eigenvalues of the independent commuting X variables.

$$[X_i, X_j] = 0$$
$$[P_i, P_j] = 0$$
$$[X_i, P_j] = i\delta_{ij} .$$

In particular, this means that a system of N interacting particles in 3 dimensions is described by one vector whose components in a basis where all the X are diagonal is a mathematical function of 3N-dimensional space describing all their possible positions, effectively a much bigger collection of values than the mere collection of N three-dimensional wavefunctions in one physical space. Schrödinger came to the same conclusion independently, and eventually proved the equivalence of his own formalism to Heisenberg's.

Since the wavefunction is a property of the whole system, not of any one part, the description in quantum mechanics is not entirely local. The description of several quantum particles has them correlated, or entangled. This entanglement leads to strange correlations between distant particles which violate the classical Bell's inequality.

Even if the particles can only be in just two positions, the wavefunction for N particles requires 2^N complex numbers, one for each total configuration of positions. This is exponentially many numbers in N, so simulating quantum mechanics on a computer requires exponential resources. Conversely, this suggests that it might be possible to find quantum systems of size N which physically compute the answers to problems which classically require 2^N bits to solve. This is the aspiration behind quantum computing.

Ehrenfest Theorem

For the time-independent operators X and P, $\partial A/\partial t = 0$ so the Heisenberg equation above reduces to:

$$i\hbar\frac{dA}{dt} = [A, H] = AH - HA,$$

where the square brackets [,] denote the commutator. For a Hamiltonian which is $p^2/2m + V(x)$, the X and P operators satisfy:

$$\frac{dX}{dt} = \frac{P}{m}, \quad \frac{dP}{dt} = -\nabla V,$$

where the first is classically the velocity, and second is classically the force, or potential gradient. These reproduce Hamilton's form of Newton's laws of motion. In the Heisenberg picture, the X and P operators satisfy the classical equations of motion. You can take the expectation value of both sides of the equation to see that, in any state $|\psi\rangle$:

$$\frac{d}{dt}\langle X\rangle = \frac{d}{dt}\langle\psi|X|\psi\rangle = \frac{1}{m}\langle\psi|P|\psi\rangle = \frac{1}{m}\langle P\rangle$$
$$\frac{d}{dt}\langle P\rangle = \frac{d}{dt}\langle\psi|P|\psi\rangle = \langle\psi|(-\nabla V)|\psi\rangle = -\langle\nabla V\rangle .$$

So Newton's laws are exactly obeyed by the expected values of the operators in any given state. This is Ehrenfest's theorem, which is an obvious corollary of the Heisenberg equations of motion, but is less trivial in the Schrödinger picture, where Ehrenfest discovered it.

Transformation Theory

In classical mechanics, a canonical transformation of phase space coordinates is one which preserves the structure of the Poisson brackets. The new variables x',p' have the same Poisson brackets with each other as the original variables x,p. Time evolution is a canonical transformation, since the phase space at any time is just as good a choice of variables as the phase space at any other time.

The Hamiltonian flow is the canonical transformation:

$$x \to x + dx = x + \frac{\partial H}{\partial p} dt$$
$$p \to p + dp = p - \frac{\partial H}{\partial x} dt\,.$$

Since the Hamiltonian can be an arbitrary function of x and p, there are such infinitesimal canonical transformations corresponding to every classical quantity G, where G serves as the Hamiltonian to generate a flow of points in phase space for an increment of time s,

$$dx = \frac{\partial G}{\partial p} ds = \{G, X\} ds$$
$$dp = -\frac{\partial G}{\partial x} ds = \{G, P\} ds\,.$$

For a general function $A(x, p)$ on phase space, its infinitesimal change at every step ds under this map is

$$dA = \frac{\partial A}{\partial x} dx + \frac{\partial A}{\partial p} dp = \{A, G\} ds\,.$$

The quantity G is called the *infinitesimal generator* of the canonical transformation.

In quantum mechanics, the quantum analog G is now a Hermitian matrix, and the equations of motion are given by commutators,

$$dA = i[G, A] ds\,.$$

The infinitesimal canonial motions can be formally integrated, just as the Heisenberg equation of motion were integrated,

$$A' = U^\dagger A U$$

where U= e^{iGs} and s is an arbitrary parameter.

The definition of a quantum canonical transformation is thus an arbitrary unitary change of basis on the space of all state vectors. U is an arbitrary unitary matrix, a complex rotation in phase space,

$$U^\dagger = U^{-1}\,.$$

These transformations leave the sum of the absolute square of the wavefunction components invariant, while they take states which are multiples of each other (including states which are imaginary multiples of each other) to states which are the same multiple of each other.

The interpretation of the matrices is that they act as generators of motions on the space of states.

For example, the motion generated by P can be found by solving the Heisenberg equation of motion using P as a Hamiltonian,

$$dX = i[X, P]ds = ds$$
$$dP = i[P, P]ds = 0 .$$

These are translations of the matrix X by a multiple of the identity matrix,

$$X \to X + sI .$$

This is the interpretation of the derivative operator D: $e^{iPs} = e^{D}$, the exponential of a derivative operator is a translation (so Lagrange's shift operator).

The X operator likewise generates translations in P. The Hamiltonian generates translations in time, the angular momentum generates rotations in physical space, and the operator $X^2 + P^2$ generates rotations in phase space.

When a transformation, like a rotation in physical space, commutes with the Hamiltonian, the transformation is called a symmetry (behind a degeneracy) of the Hamiltonian—the Hamiltonian expressed in terms of rotated coordinates is the same as the original Hamiltonian. This means that the change in the Hamiltonian under the infinitesimal symmetry generator L vanishes,

$$\frac{dH}{ds} = i[L, H] = 0 .$$

It then follows that the change in the generator under time translation also vanishes,

$$\frac{dL}{dt} = i[H, L] = 0$$

so that the matrix L is constant in time: it is conserved.

The one-to-one association of infinitesimal symmetry generators and conservation laws was discovered by Emmy Noether for classical mechanics, where the commutators are Poisson brackets, but the quantum-mechanical reasoning is identical. In quantum mechanics, any unitary symmetry transformation yields a conservation law, since if the matrix U has the property that and that the time derivative of U is zero—it is conserved.

$$U^{-1}HU = H$$

so it follows that

$$UH = HU$$

The eigenvalues of unitary matrices are pure phases, so that the value of a unitary conserved quantity is a complex number of unit magnitude, not a real number. Another way of saying this is that a unitary matrix is the exponential of i times a Hermitian matrix, so that the additive conserved real quantity, the phase, is only well-defined up to an integer multiple of 2π. Only when the unitary symmetry matrix is part of a family that comes arbitrarily close to the identity are the conserved real quantities single-valued, and then the demand that they are conserved become a much more exacting constraint.

Symmetries which can be continuously connected to the identity are called continuous, and translations, rotations, and boosts are examples. Symmetries which cannot be continuously connected to the identity are discrete, and the operation of space-inversion, or parity, and charge conjugation are examples.

The interpretation of the matrices as generators of canonical transformations is due to Paul Dirac. The correspondence between symmetries and matrices was shown by Eugene Wigner to be complete, if antiunitary matrices which describe symmetries which include time-reversal are included.

Selection Rules

It was physically clear to Heisenberg that the absolute squares of the matrix elements of X, which are the Fourier coefficients of the oscillation, would yield the rate of emission of electromagnetic radiation.

In the classical limit of large orbits, if a charge with position X(t) and charge q is oscillating next to an equal and opposite charge at position 0, the instantaneous dipole moment is q X(t), and the time variation of this moment translates directly into the space-time variation of the vector potential, which yields nested outgoing spherical waves.

For atoms, the wavelength of the emitted light is about 10,000 times the atomic radius, and the dipole moment is the only contribution to the radiative field, while all other details of the atomic charge distribution can be ignored.

Ignoring back-reaction, the power radiated in each outgoing mode is a sum of separate contributions from the square of each independent time Fourier mode of d,

$$P(\omega) = \frac{2}{3}\omega^4 |d_i|^2 .$$

Now, in Heisenberg's representation, the Fourier coefficients of the dipole moment are the matrix elements of X. This correspondence allowed Heisenberg to provide the rule for the transition intensities, the fraction of the time that, starting from an initial state i, a photon is emitted and the atom jumps to a final state j,

$$P_{ij} = \frac{2}{3}(E_i - E_j)^4 |X_{ij}|^2 .$$

This then allowed the magnitude of the matrix elements to be interpreted statistically: they give the intensity of the spectral lines, the probability for quantum jumps from the emission of dipole radiation.

Since the transition rates are given by the matrix elements of X, wherever X_{ij} is zero, the corresponding transition should be absent. These were called the selection rules, which were a puzzle until the advent of matrix mechanics.

An arbitrary state of the Hydrogen atom, ignoring spin, is labelled by $|n;\ell,m\rangle$, where the value of ℓ is a measure of the total orbital angular momentum and m is its z-component, which defines the orbit orientation. The components of the angular momentum pseudovector are

$$L_i = \epsilon_{ijk} X^j P^k$$

where the products in this expression are independent of order and real, because different components of ***X*** and ***P*** commute.

The commutation relations of ***L*** with all three coordinate matrices *X, Y, Z* (or with any vector) are easy to find,

$$[L_i, X_j] = i\epsilon_{ijk} X_k \ ,$$

which confirms that the operator ***L*** generates rotations between the three components of the vector of coordinate matrices ***X***.

From this, the commutator of L_z and the coordinate matrices *X, Y, Z* can be read off,

$$[L_z, X] = iY \ ,$$
$$[L_z, Y] = -iX \ .$$

Just like the matrix elements of X + iP and X – iP for the harmonic oscillator Hamiltonian, this commutation law implies that these operators only have certain off diagonal matrix elements in states of definite m,

$$L_z((X + iY)|m\rangle) = (X + iY)L_z|m\rangle + (X + iY)|m\rangle = (m + 1)(X + iY)|m\rangle$$

meaning that the matrix (X + iY) takes an eigenvector of L_z with eigenvalue m to an eigenvector with eigenvalue m + 1. Similarly, (X− iY)decrease m by one unit, while Z does not change the value of m.

So, in a basis of |ℓ,m⟩ states where $\boldsymbol{L}^2$ and L_z have definite values, the matrix elements of any of the three components of the position are zero, except when m is the same or changes by one unit.

This places a constraint on the change in total angular momentum. Any state can be rotated so that its angular momentum is in the z-direction as much as possible, where m = ℓ. The matrix element of the position acting on |ℓ,m⟩ can only produce values of m which are bigger by one unit, so that if the coordinates are rotated so that the final state is |ℓ',ℓ' ⟩, the value of ℓ' can be at most one bigger than the biggest value of ℓ that occurs in the initial state. So ℓ' is at most ℓ + 1.

The matrix elements vanish for ℓ' > ℓ + 1, and the reverse matrix element is determined by Hermiticity, so these vanish also when ℓ' < ℓ - 1: Dipole transitions are forbidden with a change in angular momentum of more than one unit.

Sum Rules

The Heisenberg equation of motion determines the matrix elements of P in the Heisenberg basis from the matrix elements of X.

$$P_{ij} = m\frac{d}{dt}X_{ij} = im(E_i - E_j)X_{ij}\,,$$

which turns the diagonal part of the commutation relation into a sum rule for the magnitude of the matrix elements:

$$\sum_j P_{ij}x_{ji} - X_{ij}p_{ji} = i\sum_j 2m(E_i - E_j)|X_{ij}|^2 = i\,.$$

This yields a relation for the sum of the spectroscopic intensities to and from any given state, although to be absolutely correct, contributions from the radiative capture probability for unbound scattering states must be included in the sum:

$$\sum_j 2m(E_i - E_j)|X_{ij}|^2 = 1\,.$$

REFERENCES

1. Albert Messiah (1966). Quantum Mechanics, North Holland, John Wiley & Sons. ISBN 0486409244 ; J. J. Sakurai (1994). Modern Quantum Mechanics (Addison-Wesley) ISBN 9780201539295 .
2. J. W. Negele, H. Orland (1988), Quantum Many-particle Systems, ISBN 0738200522
3. The Interaction Picture, lecture notes from New York University
4. Sakukrai, J. J.; Napolitano, Jim (2010), Modern Quantum Mechanics (2nd ed.), Addison-Wesley, ISBN 978-0805382914
5. Herbert S. Green (1965), "Matrix mechanics" (P. Noordhoff Ltd, Groningen, Netherlands) ASIN : B0006BMIP8.
6. Pauli, W (1926). "Über das Wasserstoffspektrum vom Standpunkt der neuen Quantenmechanik. Zeitschrift für Physik. 36 (5): 336–363. Bibcode:1926ZPhy...36..336P. doi:10.1007/BF01450175.
7. W. Heisenberg, Der Teil und das Ganze, Piper, Munich, (1969)The Birth of Quantum Mechanics.
8. The Birth of Quantum Mechanics
9. W. Heisenberg, Über quantentheoretische Umdeutung kinematischer und mechanischer Beziehungen, Zeitschrift für Physik, 33, 879-893, 1925 (received July 29, 1925). [English translation in: B. L. van der Waerden, editor, Sources of Quantum Mechanics (Dover Publications, 1968) ISBN 0-486-61881-1 (English title: Quantum-Theoretical Re-interpretation of Kinematic and Mechanical Relations).]
10. H. A. Kramers und W. Heisenberg, Über die Streuung von Strahlung durch Atome, Zeitschrift für Physik 31, 681-708 (1925).
11. Emilio Segrè, From X-Rays to Quarks: Modern Physicists and their Discoveries (W. H. Freeman and Company, 1980) ISBN 0-7167-1147-8, pp 153–157.
12. Abraham Pais, Niels Bohr's Times in Physics, Philosophy, and Polity (Clarendon Press, 1991) ISBN 0-19-852049-2, pp 275–279.

13. Max Born – Nobel Lecture (1954)

14. M. Born and P. Jordan, Zur Quantenmechanik, Zeitschrift für Physik, 34, 858-888, 1925 (received September 27, 1925). English translation in: B. L. van der Waerden, editor, Sources of Quantum Mechanics (Dover Publications, 1968) ISBN 0-486-61881-1

15. M. Born, W. Heisenberg, and P. Jordan, Zur Quantenmechanik II, Zeitschrift für Physik, 35, 557-615, 1925 (received November 16, 1925). English translation in: B. L. van der Waerden, editor, Sources of Quantum Mechanics (Dover Publications, 1968) ISBN 0-486-61881-1

16. Jeremy Bernstein Max Born and the Quantum Theory, Am. J. Phys.73 (11) 999-1008 (2005)

17. Mehra, Volume 3 (Springer, 2001)

18. Jammer, 1966, pp. 206-207.

19. van der Waerden, 1968, p. 51.

20. The citation by Born was in Born and Jordan's paper, the second paper in the trilogy which launched the matrix mechanics formulation. See van der Waerden, 1968, p. 351.

21. Constance Ried Courant (Springer, 1996) p. 93.

22. John von Neumann Allgemeine Eigenwerttheorie Hermitescher Funktionaloperatoren, Mathematische Annalen 102 49–131 (1929)

23. When von Neumann left Göttingen in 1932, his book on the mathematical foundations of quantum mechanics, based on Hilbert's mathematics, was published under the title Mathematische Grundlagen der Quantenmechanik. See: Norman Macrae, John von Neumann: The Scientific Genius Who Pioneered the Modern Computer, Game Theory, Nuclear Deterrence, and Much More (Reprinted by the American Mathematical Society, 1999) and Constance Reid, Hilbert (Springer-Verlag, 1996) ISBN 0-387-94674-8.

24. Bernstein, 2004, p. 1004.

25. Greenspan, 2005, p. 190.

26. Nobel Prize in Physics and 1933 – Nobel Prize Presentation Speech.

27. Bernstein, 2005, p. 1004.

28. Bernstein, 2005, p. 1006.

29. Greenspan, 2005, p. 191.

30. Greenspan, 2005, pp. 285-286.

31. Quantum Mechanics, E. Abers, Pearson Ed., Addison Wesley, Prentice Hall Inc, 2004, ISBN 978-0-13-146100-0

32. Dirac, P. A. M. (1981). The Principles of Quantum Mechanics (4th revised ed.). New York: Oxford University Press. ISBN 0-19-852011-5.

CHAPTER 5

Equations in Quantum Mechanics

Chapter Contents

Quantum mechanics, science dealing with the behaviour of matter and light on the atomic and subatomic scale. It attempts to describe and account for the properties of molecules and atoms and their constituents—electrons, protons, neutrons, and other more esoteric particles such as quarks and gluons. These properties include the interactions of the particles with one another and with electromagnetic radiation (i.e., light, X-rays, and gamma rays).

INTRODUCTION

In particle physics, the Dirac equation is a relativistic wave equation derived by British physicist Paul Dirac in 1928. In its free form, or including electromagnetic interactions, it describes all spin-1/2 massive particles such as electrons and quarks for which parity is a symmetry. It is consistent with both the principles of quantum mechanics and the theory of special relativity, and was the first theory to account fully for special relativity in the context of quantum mechanics. It was validated by accounting for the fine details of the hydrogen spectrum in a completely rigorous way.

The equation also implied the existence of a new form of matter, antimatter, previously unsuspected and unobserved and which was experimentally confirmed several years later.

It also provided a theoretical justification for the introduction of several component wave functions in Pauli's phenomenological theory of spin; the wave functions in the Dirac theory are vectors of four complex numbers (known as bispinors), two of which resemble the Pauli wavefunction in the non-relativistic limit, in contrast to the Schrödinger equation which described wave functions of only one complex value. Moreover, in the limit of zero mass, the Dirac equation reduces to the Weyl equation.

Although Dirac did not at first fully appreciate the importance of his results, the entailed explanation of spin as a consequence of the union of quantum mechanics and relativity—and the eventual discovery of the positron—represents one of the great triumphs of theoretical physics. This accomplishment has been described as fully on a par with the works of Newton, Maxwell, and Einstein before him. In the context of quantum field theory, the Dirac equation is reinterpreted to describe quantum fields corresponding to spin-1/2 particles.

Mathematical Formulation

The Dirac equation in the form originally proposed by Dirac is:

Dirac equation *(original)*

$$\left(\beta mc^2 + c\left(\sum_{n=1}^{3}\alpha_n p_n\right)\right)\psi(x,t) = i\hbar\frac{\partial\psi(x,t)}{\partial t}$$

where $\psi = \psi(x, t)$ is the wave function for the electron of rest mass m with spacetime coordinates x, t. The p_1, p_2, p_3 are the components of the momentum, understood to be the momentum operator in the Schrödinger equation. Also, c is the speed of light, and ħ is the Planck constant divided by 2π. These fundamental physical constants reflect special relativity and quantum mechanics, respectively.

Dirac's purpose in casting this equation was to explain the behavior of the relativistically moving electron, and so to allow the atom to be treated in a manner consistent with relativity. His rather modest hope was that the corrections introduced this way might have a bearing on the problem of atomic spectra. Up until that time, attempts to make the old quantum theory of the atom compatible with the theory of relativity, attempts based on discretizing the angular momentum stored in the electron's possibly non-circular orbit of the atomic nucleus, had failed – and the new quantum mechanics of Heisenberg, Pauli, Jordan, Schrödinger, and Dirac himself had not developed sufficiently to treat this problem. Although Dirac's original intentions were satisfied, his equation had far deeper implications for the structure of matter and introduced new mathematical classes of objects that are now essential elements of fundamental physics.

The new elements in this equation are the 4 × 4 matrices α_k and β, and the four-component wave function ψ. There are four components in ψbecause the evaluation of it at any given point in configuration space is a bispinor. It is interpreted as a superposition of a spin-up electron, a spin-down electron, a spin-up positron, and a spin-down positron.

The 4 × 4 matrices α_k and β are all Hermitian and have squares equal to the identity matrix:

$$\alpha_i^2 = \beta^2 = I_4$$

and they all mutually anticommute (if *i* and *j* are distinct):

$$\alpha_i\alpha_j + \alpha_j\alpha_i = 0$$
$$\alpha_i\beta + \beta\alpha_i = 0$$

The single symbolic equation thus unravels into four coupled linear first-order partial differential equations for the four quantities that make up the wave function. These matrices and the form of the wave function have a deep mathematical significance. The algebraic structure represented by the gamma matrices had been created some 50 years earlier by the English mathematician W. K. Clifford. In turn, Clifford's ideas had emerged from the mid-19th-century work of the German mathematician Hermann Grassmann in his Lineale Ausdehnungslehre (Theory of Linear Extensions). The latter had been regarded as well-nigh incomprehensible by most of his contemporaries. The appearance of something so seemingly abstract, at such a late date, and in such a direct physical manner, is one of the most remarkable chapters in the history of physics.

Making The Schrödinger Equation Relativistic

The Dirac equation is superficially similar to the Schrödinger equation for a massive free particle:

$$-\frac{\hbar^2}{2m}\nabla^2\phi = i\hbar\frac{\partial}{\partial t}\phi.$$

The left side represents the square of the momentum operator divided by twice the mass, which is the non-relativistic kinetic energy. Because relativity treats space and time as a whole, a relativistic generalization of this equation requires that space and time derivatives must enter symmetrically as they do in the Maxwell equations that govern the behavior of light — the equations must be differentially of the same order in space and time. In relativity, the momentum and the energies are the space and time parts of a spacetime vector, the four-momentum, and they are related by the relativistically invariant relation

$$\frac{E^2}{c^2} - p^2 = m^2c^2$$

which says that the length of this four-vector is proportional to the rest mass m. Substituting the operator equivalents of the energy and momentum from the Schrödinger theory, we get the Klein-Gordon equation describing the propagation of waves, constructed from relativistically invariant objects, $\left(-\frac{1}{c^2}\frac{\partial^2}{\partial t^2} + \nabla^2\right)\phi = \frac{m^2c^2}{\hbar^2}\phi$

with the wave function ϕ being a relativistic scalar: a complex number which has the same numerical value in all frames of reference. Space and time derivatives both enter to second order. This has a telling consequence for the interpretation of the equation. Because the equation is second order in the time derivative, one must specify initial values both of the wave function itself and of its first-time derivative in order to solve definite problems. Since both may be specified more or less arbitrarily, the wave function cannot maintain its former role of determining the probability density of finding the electron in a given state of motion. In the Schrödinger theory, the probability density is given by the positive definite expression

$$\rho = \phi^* \phi$$

and this density is convected according to the probability current vector

$$J = -\frac{i\hbar}{2m}(\phi^* \nabla \phi - \phi \nabla \phi^*)$$

with the conservation of probability current and density following from the continuity equation:

$$\nabla \cdot J + \frac{\partial \rho}{\partial t} = 0.$$

The fact that the density is positive definite and convected according to this continuity equation implies that we may integrate the density over a certain domain and set the total to 1, and this condition will be maintained by the conservation law. A proper relativistic theory with a probability density current must also share this feature. Now, if we wish to maintain the notion of a convected density, then we must generalize the Schrödinger expression of the density and current so that space and time derivatives again enter symmetrically in relation to the scalar wave function. We are allowed to keep the Schrödinger expression for the current, but must replace the probability density by the symmetrically formed expression

$$\rho = \frac{i\hbar}{2m}(\psi^* \partial_t \psi - \psi \partial_t \psi^*).$$

which now becomes the 4th component of a spacetime vector, and the entire probability 4-current density has the relativistically covariant expression

$$J^\mu = \frac{i\hbar}{2m}(\psi^* \partial^\mu \psi - \psi \partial^\mu \psi^*)$$

The continuity equation is as before. Everything is compatible with relativity now, but we see immediately that the expression for the density is no longer positive definite – the initial values of both ψ and $\partial_t \psi$ may be freely chosen, and the density may thus become negative, something that is impossible for a legitimate probability density. Thus, we cannot get a simple generalization of the Schrödinger equation under the naive assumption that the wave function is a relativistic scalar, and the equation it satisfies, second order in time.

Although it is not a successful relativistic generalization of the Schrödinger equation, this equation is resurrected in the context of quantum field theory, where it is known as the Klein–Gordon equation, and describes a spinless particle field (e.g. pi meson).

Historically, Schrödinger himself arrived at this equation before the one that bears his name but soon discarded it. In the context of quantum field theory, the indefinite density is understood to correspond to the charge density, which can be positive or negative, and not the probability density.

Dirac's Coup

Dirac thus thought to try an equation that was first order in both space and time. One could, for example, formally (i.e. by abuse of notation) take the relativistic expression for the energy

$$E = c\sqrt{p^2 + m^2c^2},$$

replace p by its operator equivalent, expand the square root in an infinite series of derivative operators, set up an eigenvalue problem, then solve the equation formally by iterations. Most physicists had little faith in such a process, even if it was technically possible.

As the story goes, Dirac was staring into the fireplace at Cambridge, pondering this problem, when he hit upon the idea of taking the square root of the wave operator thus:

$$\nabla^2 - \frac{1}{c^2}\frac{\partial^2}{\partial t^2} = \left(A\partial_x + B\partial_y + C\partial_z + \frac{i}{c}D\partial_t\right)\left(A\partial_x + B\partial_y + C\partial_z + \frac{i}{c}D\partial_t\right).$$

On multiplying out the right side we see that, in order to get all the cross-terms such as $\partial_x\partial_y$ to vanish, we must assume

$$AB + BA = 0, \ldots$$

with

$$A^2 = B^2 = \ldots = 1.$$

Dirac, who had just then been intensely involved with working out the foundations of Heisenberg's matrix mechanics, immediately understood that these conditions could be met if A, B, C and D are matrices, with the implication that the wave function has multiple components. This immediately explained the appearance of two-component wave functions in Pauli's phenomenological theory of spin, something that up until then had been regarded as mysterious, even to Pauli himself. However, one needs at least 4 × 4 matrices to set up a system with the properties required — so the wave function had four components, not two, as in the Pauli theory, or one, as in the bare Schrödinger theory. The four-component wave function represents a new class of mathematical object in physical theories that makes its first appearance here.

Given the factorization in terms of these matrices, one can now write down immediately an equation

$$\left(A\partial_x + B\partial_y + C\partial_z + \frac{i}{c}D\partial_t\right)\psi = \kappa\psi$$

with κ to be determined. Applying again the matrix operator on both sides yields

$$\left(\nabla^2 - \frac{1}{c^2}\partial_t^2\right)\psi = \kappa^2\psi.$$

On taking $\kappa = mc/\hbar$ we find that all the components of the wave function *individually* satisfy the relativistic energy–momentum relation. Thus the sought-for equation that is first-order in both space and time is

$$\left(A\partial_x + B\partial_y + C\partial_z + \frac{i}{c}D\partial_t - \frac{mc}{\hbar}\right)\psi = 0.$$

Setting

$$A = i\beta\alpha_1\,,\; B = i\beta\alpha_2\,,\; C = i\beta\alpha_3\,,\; D = \beta\,,$$

and because $D^2 = \beta^2 = I_4\,,$

we get the Dirac equation as written above.

Covariant form and Relativistic Invariance

To demonstrate the relativistic invariance of the equation, it is advantageous to cast it into a form in which the space and time derivatives appear on an equal footing. New matrices are introduced as follows:

$$D = \gamma^0\,,$$
$$A = i\gamma^1\,,\; B = i\gamma^2\,,\; C = i\gamma^3\,,$$

and the equation takes the form (remembering the definition of the covariant components of the 4-gradient and especially that $\partial_0 = \frac{1}{c}\partial_t$)

Dirac equation

$$i\hbar\gamma^\mu\partial_\mu\psi - mc\psi = 0$$

where there is an implied summation over the values of the twice-repeated index μ = 0, 1, 2, 3, and ∂_μ is the 4-gradient. In practice one often writes the gamma matrices in terms of 2 × 2 sub-matrices taken from the Pauli matrices and the 2 × 2 identity matrix. Explicitly the standard representation is

$$\gamma^0 = \begin{pmatrix} I_2 & 0 \\ 0 & -I_2 \end{pmatrix},\gamma^1 = \begin{pmatrix} 0 & \sigma_x \\ -\sigma_x & 0 \end{pmatrix},\gamma^2 = \begin{pmatrix} 0 & \sigma_y \\ -\sigma_y & 0 \end{pmatrix},\gamma^3 = \begin{pmatrix} 0 & \sigma_z \\ -\sigma_z & 0 \end{pmatrix}.$$

The complete system is summarized using the Minkowski metric on spacetime in the form

$$\{\gamma^\mu, \gamma^\nu\} = 2\eta^{\mu\nu} I_4$$

where the bracket expression

$$\{a, b\} = ab + ba$$

denotes the anticommutator. These are the defining relations of a Clifford algebra over a pseudo-orthogonal 4-dimensional space with metric signature (+ − − −). The specific Clifford algebra employed in the Dirac equation is known today as the Dirac algebra.

Although not recognized as such by Dirac at the time the equation was formulated, in hindsight the introduction of this geometric algebra represents an enormous stride forward in the development of quantum theory.

The Dirac equation may now be interpreted as an eigenvalue equation, where the rest mass is proportional to an eigenvalue of the 4-momentum operator, the proportionality constant being the speed of light:

$$P_{\mathrm{op}}\psi = mc\psi.$$

Using $\partial\!\!\!/$ (pronounced: "d-slash"[4]) in Feynman slash notation, which includes the gamma matrices as well as a summation over the spinor components in the derivative itself, the Dirac equation becomes:

$$i\hbar\partial\!\!\!/\psi - mc\psi = 0$$

In practice, physicists often use units of measure such that $\hbar = c = 1$, known as natural units. The equation then takes the simple form

Dirac equation *(natural units)*

$$(i\partial\!\!\!/ - m)\psi = 0$$

A fundamental theorem states that if two distinct sets of matrices are given that both satisfy the Clifford relations, then they are connected to each other by a similarity transformation:

$$\gamma^{\mu\prime} = S^{-1}\gamma^\mu S.$$

If in addition the matrices are all unitary, as are the Dirac set, then S itself is unitary;

$$\gamma^{\mu\prime} = U^\dagger\gamma^\mu U.$$

The transformation U is unique up to a multiplicative factor of absolute value 1. Let us now imagine a Lorentz transformation to have been performed on the space and time coordinates, and on the derivative operators, which form a covariant vector. For the operator $\gamma^\mu\partial_\mu$ to remain invariant, the gammas must transform among themselves as a contravariant vector with respect to their spacetime index. These new gammas will themselves satisfy the Clifford relations, because of the orthogonality of the Lorentz transformation. By the fundamental theorem, we may replace the new set by the old set subject to a unitary transformation. In the new frame, remembering that the rest mass is a relativistic scalar, the Dirac equation will then take the form

$$(iU^\dagger\gamma^\mu U\partial'_\mu - m)\psi(x',t') = 0$$
$$U^\dagger(i\gamma^\mu\partial'_\mu - m)U\psi(x',t') = 0.$$

If we now define the transformed spinor

$$\psi' = U\psi$$

then we have the transformed Dirac equation in a way that demonstrates manifest relativistic invariance:

$$(i\gamma^{\mu}\partial'_{\mu} - m)\psi'(x',t') = 0.$$

Thus, once we settle on any unitary representation of the gammas, it is final provided we transform the spinor according to the unitary transformation that corresponds to the given Lorentz transformation. The various representations of the Dirac matrices employed will bring into focus particular aspects of the physical content in the Dirac wave function. The representation shown here is known as the standard representation – in it, the wave function's upper two components go over into Pauli's 2-spinor wave function in the limit of low energies and small velocities in comparison to light.

The considerations above reveal the origin of the gammas in geometry, hearkening back to Grassmann's original motivation – they represent a fixed basis of unit vectors in spacetime. Similarly, products of the gammas such as $\gamma_{\mu}\gamma_{\nu}$ represent oriented surface elements, and so on. With this in mind, we can find the form of the unit volume element on spacetime in terms of the gammas as follows. By definition, it is

$$V = \frac{1}{4!}\epsilon_{\mu\nu\alpha\beta}\gamma^{\mu}\gamma^{\nu}\gamma^{\alpha}\gamma^{\beta}.$$

For this to be an invariant, the epsilon symbol must be a tensor, and so must contain a factor of $\sqrt{g}$, where g is the determinant of the metric tensor. Since this is negative, that factor is *imaginary*. Thus

$$V = i\gamma^0\gamma^1\gamma^2\gamma^3.$$

This matrix is given the special symbol γ^5, owing to its importance when one is considering improper transformations of spacetime, that is, those that change the orientation of the basis vectors. In the standard representation, it is

$$\gamma_5 = \begin{pmatrix} 0 & I_2 \\ I_2 & 0 \end{pmatrix}.$$

This matrix will also be found to anticommute with the other four Dirac matrices:

$$\gamma^5\gamma^{\mu} + \gamma^{\mu}\gamma^5 = 0$$

It takes a leading role when questions of parity arise because the volume element as a directed magnitude changes sign under a spacetime reflection. Taking the positive square root above thus amounts to choosing a handedness convention on spacetime .

Conservation of Probability Current

By defining the adjoint spinor

$$\bar{\psi} = \psi^{\dagger}\gamma^0$$

where $\psi^{\dagger}$ is the conjugate transpose of ψ, and noticing that

$$(\gamma^{\mu})^{\dagger}\gamma^0 = \gamma^0\gamma^{\mu}.$$

we obtain, by taking the Hermitian conjugate of the Dirac equation and multiplying from the right by γ^0, the adjoint equation:

$$\bar{\psi}(-i\gamma^{\mu}\partial_{\mu} - m) = 0$$

where ∂_{μ} is understood to act to the left. Multiplying the Dirac equation by ψ from the left, and the adjoint equation by ψ from the right, and subtracting, produces the law of conservation of the Dirac current:

$$\partial_{\mu}\left(\bar{\psi}\gamma^{\mu}\psi\right) = 0.$$

Now we see the great advantage of the first-order equation over the one Schrödinger had tried – this is the conserved current density required by relativistic invariance, only now its 4th component is *positive definite* and thus suitable for the role of a probability density:

$$J^0 = \bar{\psi}\gamma^0\psi = \psi^{\dagger}\psi.$$

Because the probability density now appears as the fourth component of a relativistic vector and not a simple scalar as in the Schrödinger equation, it will be subject to the usual effects of the Lorentz transformations such as time dilation. Thus, for example, atomic processes that are observed as rates, will necessarily be adjusted in a way consistent with relativity, while those involving the measurement of energy and momentum, which themselves form a relativistic vector, will undergo parallel adjustment which preserves the relativistic covariance of the observed values.

Solutions

Dirac spinor for details of solutions to the Dirac equation. Note that since the Dirac operator acts on 4-tuples of square-integrable functions, its solutions should be members of the same Hilbert space. The fact that the energies of the solutions do not have a lower bound is unexpected – see the hole theory section below for more details.

Comparison with the PAULI Theory

The necessity of introducing half-integer spin goes back experimentally to the results of the Stern–Gerlach experiment. A beam of atoms is run through a strong inhomogeneous magnetic field, which then splits into N parts depending on the intrinsic angular momentum of the atoms. It was found that for silver atoms, the beam was split in two—the ground state therefore could not be integer, because even if the intrinsic angular momentum of the atoms were as small as possible, 1, the beam would be split into three parts, corresponding to atoms with $L_z = -1, 0, +1$. The conclusion is that silver atoms have net intrinsic angular momentum of $\frac{1}{2}$. Pauli set up a theory which explained this splitting by introducing a two-component wave function and a corresponding correction term in the Hamiltonian, representing a semi-classical coupling of this wave function to an applied magnetic field, as so in SI units: (Note that bold faced characters imply Euclidean vectors in 3 dimensions, where as the Minkowski four-vector A_{μ} can be defined as $A_{\mu} = (\Phi/c, -\mathbf{A})$.)

$$H = \frac{1}{2m}(\boldsymbol{\sigma}\cdot(\mathbf{p} - e\mathbf{A}))^2 + e\phi.$$

Here **A** and ϕ represent the components of the electromagnetic four-potential in their standard SI units, and the three sigmas are the Pauli matrices. On squaring out the first term, a residual interaction with the magnetic field is found, along with the usual classical Hamiltonian of a charged particle interacting with an applied field in SI units:

$$H = \frac{1}{2m}(\mathbf{p} - e\mathbf{A})^2 + e\phi - \frac{e\hbar}{2m}\boldsymbol{\sigma}\cdot\mathbf{B}.$$

This Hamiltonian is now a 2 × 2 matrix, so the Schrödinger equation based on it must use a two-component wave function. Pauli had introduced the 2 × 2 sigma matrices as pure phenomenology— Dirac now had a theoretical argument that implied that spin was somehow the consequence of the marriage of quantum mechanics to relativity. On introducing the external electromagnetic 4-vector potential into the Dirac equation in a similar way, known as minimal coupling, it takes the form :

$$(\gamma^\mu(i\hbar\partial_\mu - \frac{e}{c}A_\mu) - mc)\psi = 0\,.$$

A second application of the Dirac operator will now reproduce the Pauli term exactly as before, because the spatial Dirac matrices multiplied by i, have the same squaring and commutation properties as the Pauli matrices. What is more, the value of the gyromagnetic ratio of the electron, standing in front of Pauli's new term, is explained from first principles. This was a major achievement of the Dirac equation and gave physicists great faith in its overall correctness. There is more however. The Pauli theory may be seen as the low energy limit of the Dirac theory in the following manner. First the equation is written in the form of coupled equations for 2-spinors with the SI units restored:

$$\begin{pmatrix} (mc^2 - E + e\phi) & c\boldsymbol{\sigma}\cdot(\mathbf{p} - e\mathbf{A}) \\ -c\boldsymbol{\sigma}\cdot(\mathbf{p} - e\mathbf{A}) & (mc^2 + E - e\phi) \end{pmatrix} \begin{pmatrix} \psi_+ \\ \psi_- \end{pmatrix} = \begin{pmatrix} 0 \\ 0 \end{pmatrix}.$$

so

$$(E - e\phi)\psi_+ - c\boldsymbol{\sigma}\cdot(\mathbf{p} - e\mathbf{A})\,\psi_- = mc^2\psi_+$$
$$-(E - e\phi)\psi_- + c\boldsymbol{\sigma}\cdot(\mathbf{p} - e\mathbf{A})\,\psi_+ = mc^2\psi_-$$

Assuming the field is weak and the motion of the electron non-relativistic, we have the total energy of the electron approximately equal to its rest energy, and the momentum going over to the classical value,

$$E - e\phi \approx mc^2$$
$$\mathbf{p} \approx m\mathbf{v}$$

and so the second equation may be written

$$\psi_- \approx \frac{1}{2mc}\boldsymbol{\sigma}\cdot(\mathbf{p} - e\mathbf{A})\,\psi_+$$

which is of order *v/c* – thus at typical energies and velocities, the bottom components of the Dirac spinor in the standard representation are much suppressed in comparison to the top components. Substituting this expression into the first equation gives after some rearrangement

$$(E - mc^2)\psi_+ = \frac{1}{2m}[\boldsymbol{\sigma} \cdot (\mathbf{p} - e\mathbf{A})]^2 \psi_+ + e\phi\psi_+$$

The operator on the left represents the particle energy reduced by its rest energy, which is just the classical energy, so we recover Pauli's theory if we identify his 2-spinor with the top components of the Dirac spinor in the non-relativistic approximation. A further approximation gives the Schrödinger equation as the limit of the Pauli theory. Thus, the Schrödinger equation may be seen as the far non-relativistic approximation of the Dirac equation when one may neglect spin and work only at low energies and velocities. This also was a great triumph for the new equation, as it traced the mysterious i that appears in it, and the necessity of a complex wave function, back to the geometry of spacetime through the Dirac algebra. It also highlights why the Schrödinger equation, although superficially in the form of a diffusion equation, actually represents the propagation of waves.

It should be strongly emphasized that this separation of the Dirac spinor into large and small components depends explicitly on a low-energy approximation. The entire Dirac spinor represents an irreducible whole, and the components we have just neglected to arrive at the Pauli theory will bring in new phenomena in the relativistic regime – antimatter and the idea of creation and annihilation of particles.

Comparison with the Weyl Theory

In the limit $m \to 0$, the Dirac equation reduces to the Weyl equation, which describes relativistic massless spin-$\frac{1}{2}$ particles.

Dirac Lagrangian

Both the Dirac equation and the Adjoint Dirac equation can be obtained from (varying) the action with a specific Lagrangian density that is given by:

$$\mathcal{L} = i\hbar c\overline{\psi}\gamma^\mu \partial_\mu \psi - mc^2\overline{\psi}\psi$$

If one varies this with respect to ψ one gets the Adjoint Dirac equation. Meanwhile, if one varies this with respect to ψ one gets the Dirac equation.

Physical Interpretation

Identification of Observables

The critical physical question in a quantum theory is—what are the physically observable quantities defined by the theory? According to the postulates of quantum mechanics, such quantities are defined by Hermitian operators that act on the Hilbert space of possible states of a system. The eigenvalues of these operators are then the possible results of measuring the corresponding physical quantity. In the Schrödinger theory, the simplest such object is the overall Hamiltonian, which represents the total energy of the system. If we wish to maintain this interpretation on passing to the Dirac theory, we must take the Hamiltonian to be

$$H = \gamma^0 \left[mc^2 + c\gamma^k \left(p_k - \frac{q}{c} A_k \right) \right] + qA^0.$$

where, as always, there is an implied summation over the twice-repeated index k = 1, 2, 3. This looks promising, because we see by inspection the rest energy of the particle and, in case A = 0, the energy of a charge placed in an electric potential qA^0. What about the term involving the vector potential? In classical electrodynamics, the energy of a charge moving in an applied potential is

$$H = c\sqrt{\left(p - \frac{q}{c} A \right)^2 + m^2c^2} + qA^0.$$

Thus, the Dirac Hamiltonian is fundamentally distinguished from its classical counterpart, and we must take great care to correctly identify what is observable in this theory. Much of the apparently paradoxical behavior implied by the Dirac equation amounts to a misidentification of these observables.

Hole Theory

The negative E solutions to the equation are problematic, for it was assumed that the particle has a positive energy. Mathematically speaking, however, there seems to be no reason for us to reject the negative-energy solutions. Since they exist, we cannot simply ignore them, for once we include the interaction between the electron and the electromagnetic field, any electron placed in a positive-energy eigenstate would decay into negative-energy eigenstates of successively lower energy. Real electrons obviously do not behave in this way, or they would disappear by emitting energy in the form of photons.

To cope with this problem, Dirac introduced the hypothesis, known as hole theory, that the vacuum is the many-body quantum state in which all the negative-energy electron eigenstates are occupied. This description of the vacuum as a "sea" of electrons is called the Dirac sea. Since the Pauli exclusion principle forbids electrons from occupying the same state, any additional electron would be forced to occupy a positive-energy eigenstate, and positive-energy electrons would be forbidden from decaying into negative-energy eigenstates.

If an electron is forbidden from simultaneously occupying positive-energy and negative-energy eigenstates, then the feature is known as Zitterbewegung, which arises from the interference of positive-energy and negative-energy states, would have to be considered to be an unphysical prediction of time-dependent Dirac theory. This conclusion may be inferred from the explanation of hole theory given in the preceding paragraph. Recent results have been published in Nature in which the Zitterbewegung feature was simulated in a trapped-ion experiment. This experiment impacts the hole interpretation if one infers that the physics-laboratory experiment is not merely a check on the mathematical correctness of a Dirac-equation solution but the measurement of a real effect whose detectability in electron physics is still beyond reach.

Dirac further reasoned that if the negative-energy eigenstates are incompletely filled, each unoccupied eigenstate – called a **hole** – would behave like a positively charged particle. The hole possesses a positive energy since energy is required to create a particle–hole pair from the vacuum. As noted above, Dirac initially thought that the hole might be the proton, but Hermann Weyl pointed out that the hole should behave as if it had the same mass as an electron, whereas the proton is over 1800 times heavier. The hole was eventually identified as the positron, experimentally discovered by Carl Anderson in 1932.

It is not entirely satisfactory to describe the "vacuum" using an infinite sea of negative-energy electrons. The infinitely negative contributions from the sea of negative-energy electrons have to be canceled by an infinite positive "bare" energy and the contribution to the charge density and current coming from the sea of negative-energy electrons is exactly canceled by an infinite positive "jellium" background so that the net electric charge density of the vacuum is zero. In quantum field theory, a Bogoliubov transformation on the creation and annihilation operators (turning an occupied negative-energy electron state into an unoccupied positive energy positron state and an unoccupied negative-energy electron state into an occupied positive energy positron state) allows us to bypass the Dirac sea formalism even though, formally, it is equivalent to it.

In certain applications of condensed matter physics, however, the underlying concepts of "hole theory" are valid. The sea of conduction electrons in an electrical conductor, called a Fermi sea, contains electrons with energies up to the chemical potential of the system. An unfilled state in the Fermi sea behaves like a positively charged electron, though it is referred to as a "hole" rather than a "positron". The negative charge of the Fermi sea is balanced by the positively charged ionic lattice of the material.

In Quantum Field Theory

In quantum field theories such as quantum electrodynamics, the Dirac field is subject to a process of second quantization, which resolves some of the paradoxical features of the equation.

Other Formulations

The Dirac equation can be formulated in a number of other ways.

As A Differential Equation in one Real Component

Generically (if a certain linear function of electromagnetic field does not vanish identically), three out of four components of the spinor function in the Dirac equation can be algebraically eliminated, yielding an equivalent fourth-order partial differential equation for just one component. Furthermore, this remaining component can be made real by a gauge transform.

Curved Spacetime

This article has developed the Dirac equation in flat spacetime according to special relativity. It is possible to formulate the Dirac equation in curved spacetime.

The Algebra of Physical Space

This article developed the Dirac equation using four vectors and Schrödinger operators. The Dirac equation in the algebra of physical space uses a Clifford algebra over the real numbers, a type of geometric algebra.

KLEIN–GORDON EQUATION

The Klein–Gordon equation (Klein–Fock–Gordon equation or sometimes Klein–Gordon–Fock equation) is a relativistic wave equation, related to the Schrödinger equation. It is second order in space and time and manifestly Lorentz covariant. It is a quantized version of the relativistic energy-momentum relation. Its solutions include a quantum scalar or pseudoscalar field, a field whose quanta are spinless particles. Its theoretical relevance is similar to that of the Dirac equation. Electromagnetic interactions can be incorporated, forming the topic of scalar electrodynamics, but because common spinless particles like the pi-mesons are unstable and also experience the strong interaction (with unknown interaction term in the Hamiltonian), the practical utility is limited.

The equation can be put into the form of a Schrödinger equation. In this form it is two coupled differential equations, each of first order in time. The solutions have two components, reflecting the charge degree of freedom in relativity. It admits a conserved quantity, but this is not positive definite. The wave function cannot therefore be interpreted as a probability amplitude. The conserved quantity is instead interpreted as electric charge and the norm squared of the wave function is interpreted as a charge density. The equation describes all spinless particles with positive, negative as well as zero charge. Within the Feynman–Stueckelberg interpretation, particles and antiparticles are treated mathematically as if' they propagate forward and backward in time respectively, the advanced propagator (as opposed to the retarded propagator) is employed for antiparticles. Physically, all particles move forward in time.

Any solution of the free Dirac equation is componentwise a solution of the free Klein–Gordon equation.

The equation does not form the basis for a consistent quantum relativistic one-particle theory. There is no known such theory for particles of any spin. For full reconciliation of quantum mechanics with special relativity quantum field theory is needed, in which the Klein–Gordon equation reemerges as the equation obeyed by the components of all free quantum fields. In quantum field theory, the solutions of the free (noninteracting) versions of the original equations still play a role. They are needed to build the Hilbert space (Fock space) and to express quantum field by using complete sets (spanning sets of Hilbert space) of wave functions.

The equation was named after the physicists Oskar Klein and Walter Gordon, who in 1926 proposed that it describes relativistic electrons. Other authors making similar claims in that same year were Vladimir Fock, Johann Kudar, Théophile de Donder and Frans-H. van den Dungen, and Louis de Broglie. Although it turned out that modeling the electron's spin required the Dirac equation, the Klein–Gordon equation correctly describes the spinless relativisitic composite particles, like the pion. On July 4, 2012 CERN announced the discovery of the Higgs boson. Since the Higgs boson is a spin-zero particle, it is the first observed ostensibly elementary particle to be described by the Klein–Gordon equation. Further experimentation and analysis is required to discern whether the Higgs boson observed is that of the Standard Model, or a more exotic, possibly composite, form.

The Klein–Gordon equation was first considered as a quantum wave equation by Schrödinger in his search for an equation describing de Broglie waves. The equation is found in his notebooks from late 1925, and he appears to have prepared a manuscript applying it to the hydrogen atom. Yet, because it fails to take into account the electron's spin, the equation predicts the hydrogen atom's fine structure incorrectly, including overestimating the overall magnitude of the splitting pattern by a factor of 4n/2n − 1 for the n-th energy level. The Dirac equation relativistic spectrum is, however, easily recovered if the orbital momentum quantum number ℓ is replaced by total angular momentum quantum number j. In January 1926, Schrödinger submitted for publication instead his equation, a non-relativistic approximation that predicts the Bohr energy levels of hydrogen without fine structure.

In 1926, soon after the Schrödinger equation was introduced, Vladimir Fock wrote an article about its generalization for the case of magnetic fields, where forces were dependent on velocity, and independently derived this equation. Both Klein and Fock used Kaluza and Klein's method. Fock also determined the gauge theory for the wave equation. The Klein–Gordon equation for a free particle has a simple plane wave solution.

PAULI EQUATION

In quantum mechanics, the Pauli equation or Schrödinger–Pauli equation is the formulation of the Schrödinger equation for spin-½ particles, which takes into account the interaction of the particle's spin with an external electromagnetic field. It is the non-relativistic limit of the Dirac equation and can be used where particles are moving at speeds much less than the speed of light, so that relativistic effects can be neglected. It was formulated by Wolfgang Pauli in 1927.

Equation

For a particle of mass m and charge q, in an electromagnetic field described by the vector potential $\mathbf{A} = (A_x, A_y, A_z)$ and scalar electric potential ϕ, the Pauli equation reads:

Pauli equation *(general)*

$$\left[\frac{1}{2m}(\boldsymbol{\sigma}\cdot(\mathbf{p}-q\mathbf{A}))^2 + q\phi\right]|\psi\rangle = i\hbar\frac{\partial}{\partial t}|\psi\rangle$$

where $\boldsymbol{\sigma} = (\sigma_x, \sigma_y, \sigma_z)$ are the Pauli matrices collected into a vector for convenience, $\mathbf{p} = -i\hbar\nabla$ is the momentum operator wherein ∇ denotes the gradient operator, and

$$|\psi\rangle = \begin{pmatrix}\psi_+\\ \psi_-\end{pmatrix}$$

is the two-component spinor wavefunction, a column vector written in Dirac notation.

The Hamiltonian operator

$$\hat{H} = \frac{1}{2m}(\boldsymbol{\sigma}\cdot(\mathbf{p} - q\mathbf{A}))^2 + q\phi$$

is a 2 × 2 matrix operator, because of the Pauli matrices. Substitution into the Schrödinger equation gives the Pauli equation. This Hamiltonian is similar to the classical Hamiltonian for a charged particle interacting with an electromagnetic field, see Lorentz force for details of this classical case. The kinetic energy term for a free particle in the absence of an electromagnetic field is just $\mathbf{p}^2/2m$ where **p** is the kinetic momentum, while in the presence of an EM field we have the minimal coupling $\mathbf{p} = \mathbf{P} - q\mathbf{A}$, where **P** is the canonical momentum.

The Pauli matrices can be removed from the kinetic energy term, using the Pauli vector identity:

$$(\boldsymbol{\sigma}\cdot\mathbf{a})(\boldsymbol{\sigma}\cdot\mathbf{b}) = \mathbf{a}\cdot\mathbf{b} + i\boldsymbol{\sigma}\cdot(\mathbf{a}\times\mathbf{b})$$

to obtain

Pauli equation *(standard form)*

$$\hat{H}|\psi\rangle = \left[\frac{1}{2m}\left[(\mathbf{p} - q\mathbf{A})^2 - q\hbar\boldsymbol{\sigma}\cdot\mathbf{B}\right] + q\phi\right]|\psi\rangle = i\hbar\frac{\partial}{\partial t}|\psi\rangle$$

where $\mathbf{B} = \nabla \times \mathbf{A}$ is the magnetic field.

Relationship with the Schrödinger Equation and the Dirac Equation

The Pauli equation is non-relativistic, but it does incorporate spin. As such, it can be thought of as occupying the middle ground between:

- The familiar Schrödinger equation (on a complex scalar wavefunction), which is non-relativistic and does not predict spin.
- The Dirac equation (on a complex four-component spinor), which is fully relativistic (with respect to special relativity) and predicts spin.

Note that because of the properties of the Pauli matrices, if the magnetic vector potential **A** is equal to zero, then the equation reduces to the familiar Schrödinger equation for a particle in a purely electric potential ϕ, except that it operates on a two-component spinor:

$$\left[\frac{\mathbf{p}^2}{2m} + q\phi\right]\begin{pmatrix}\psi_+\\ \psi_-\end{pmatrix} = i\hbar\frac{\partial}{\partial t}\begin{pmatrix}\psi_+\\ \psi_-\end{pmatrix}.$$

Therefore, we can see that the spin of the particle only affects its motion in the presence of a magnetic field.

Relationship With Stern–Gerlach Experiment

Both spinor components satisfy the Schrödinger equation. For a particle in an externally applied **B** field, the Pauli equation reads:

Pauli equation *(B-field)*

$$\underbrace{i\hbar\frac{\partial}{\partial t}|\psi_\pm\rangle = \left(\frac{(\mathbf{p}-q\mathbf{A})^2}{2m}+q\phi\right)\hat{1}|\psi\rangle}_{\text{Schrödinger equation}} - \underbrace{\frac{q\hbar}{2m}\sigma\cdot\mathbf{B}|\psi\rangle}_{\text{Stern-Gerlach term}}$$

where

$$\hat{1} = \begin{pmatrix} 1 & 0 \\ 0 & 1 \end{pmatrix}$$

is the 2 × 2 identity matrix, which acts as an identity operator.

The Stern–Gerlach term can obtain the spin orientation of atoms with one valence electron, e.g. silver atoms which flow through an inhomogeneous magnetic field.

Analogously, the term is responsible for the splitting of spectral lines (corresponding to energy levels) in a magnetic field as can be viewed in the anomalous Zeeman effect.

RYDBERG FORMULA

The Rydberg formula is used in atomic physics to describe the wavelengths of spectral lines of many chemical elements. It was formulated by the Swedish physicist Johannes Rydberg, and presented on 5 November 1888.

In 1880, Rydberg worked on a formula describing the relation between the wavelengths in spectral lines of alkali metals. He noticed that lines came in series and he found that he could simplify his calculations by using the wavenumber (the number of waves occupying the unit length, equal to 1/λ, the inverse of the wavelength) as his unit of measurement. He plotted the wavenumbers (n) of successive lines in each series against consecutive integers which represented the order of the lines in that particular series. Finding that the resulting curves were similarly shaped, he sought a single function which could generate all of them, when appropriate constants were inserted.

First he tried the formula: $n=n_0-\frac{C_0}{m+m'}$, where n is the line's wavenumber, n_0 is the series limit, m is the line's ordinal number in the series, m' is a constant different for different series and C_0 is a universal constant. This did not work very well.

Rydberg was trying: $n=n_0-\frac{C_0}{(m+m')^2}$ when he became aware of Balmer's formula for the hydrogen spectrum $\lambda=\frac{hm^2}{m^2-4}$ In this equation, m is an integer and h is a constant (not to be confused with the later Planck's constant).

Rydberg therefore rewrote Balmer's formula in terms of wavenumbers, as $n=n_0-\frac{4n_0}{m^2}$.

This suggested that the Balmer formula for hydrogen might be a special case with $m'=0$ and $C_0=4n_0$ where $n_0=\frac{1}{h}$, the reciprocal of Balmer's constant (this constant **h** is written **B** in the Balmer equation article, again to avoid confusion with Planck's constant).

The term C_0 was found to be a universal constant common to all elements, equal to 4/h. This constant is now known as the Rydberg constant, and m' is known as the quantum defect.

As stressed by Niels Bohr, expressing results in terms of wavenumber, not wavelength, was the key to Rydberg's discovery. The fundamental role of wavenumbers was also emphasized by the Rydberg-Ritz combination principle of 1908. The fundamental reason for this lies in quantum mechanics. Light's wavenumber is proportional to frequency $\frac{1}{\lambda}=\frac{f}{c}$, and therefore also proportional to light's quantum energy E. Thus $\frac{1}{\lambda}=\frac{E}{hc}$. Modern understanding is that Rydberg's findings were a reflection of the underlying simplicity of the behavior of spectral lines, in terms of fixed (quantized) energy differences between electron orbitals in atoms. Rydberg's 1888 classical expression for the form of the spectral series was not accompanied by a physical explanation. Ritz's pre-quantum 1908 explanation for the mechanism underlying the spectral series was that atomic electrons behaved like magnets and that the magnets could vibrate with respect to the atomic nucleus (at least temporarily) to produce electromagnetic radiation, but this theory was superseded in 1913 by Niels Bohr's model of the atom.

In Bohr's conception of the atom, the integer Rydberg (and Balmer) n numbers represent electron orbitals at different integral distances from the atom. A frequency (or spectral energy) emitted in a transition from n_1 to n_2 therefore represents the photon energy emitted or absorbed when an electron makes a jump from orbital 1 to orbital 2.

Later models found that the values for n_1 and n_2 corresponded to the principal quantum numbers of the two orbitals.

Rydberg Formula for Hydrogen

$$\frac{1}{\lambda_{\text{vac}}} = R\left(\frac{1}{n_1^2} - \frac{1}{n_2^2}\right)$$

Where

λ_{vac} is the wavelength of electromagnetic radiation emitted in vacuum,

R is the Rydberg constant, approximately $1.097 \times 10^7\ \text{m}^{-1}$,

n_1 and n_2 are integers greater than or equal to 1 such that $n_1 < n_2$, corresponding to the principal quantum numbers of the orbitals occupied before and after the 'quantum leap'.

By setting n_1 to 1 and letting n_2 run from 2 to infinity, the spectral lines known as the Lyman series converging to 91 nm are obtained, in the same manner:

n_1	n_2	Name	Converge toward
1	$2 \rightarrow \infty$	Lyman series	91.13 nm (UV)
2	$3 \rightarrow \infty$	Balmer series	364.51 nm (Visible)
3	$4 \rightarrow \infty$	Paschen series	820.14 nm (IR)
4	$5 \rightarrow \infty$	Brackett series	1458.03 nm (Far IR)

5	$6 \rightarrow \infty$	Pfund series	2278.17 nm (Far IR)
6	$7 \rightarrow \infty$	Humphreys series	3280.56 nm (Far IR)

Rydberg Formula for any Hydrogen-Like Element

The formula above can be extended for use with any hydrogen-like chemical elements with

$$\frac{1}{\lambda_{\text{vac}}} = RZ^2 \left(\frac{1}{n_1^2} - \frac{1}{n_2^2} \right)$$

where

λ_{vac} is the wavelength of the light emitted in vacuum;

R is the Rydberg constant for this element;

Z is the atomic number, i.e. the number of protons in the atomic nucleus of this element;

n_1 and n_2 are integers such that $n_1 < n_2$ corresponding to the principal quantum numbers of the orbitals occupied before and after.

It's important to notice that this formula can be directly applied only to hydrogen-like, also called hydrogenic atoms of chemical elements, i.e. atoms with only one electron being affected by an effective nuclear charge (which is easily estimated). Examples would include He^+, Li^{2+}, Be^{3+} etc., where no other electrons exist in the atom.

But the Rydberg formula also provides correct wavelengths for distant electrons, where the effective nuclear charge can be estimated as the same as that for hydrogen, since all but one of the nuclear charges have been screened by other electrons, and the core of the atom has an effective positive charge of +1.

Finally, with certain modifications (replacement of ***Z*** by ***Z*−1**, and use of the integers 1 and 2 for the ns to give a numerical value of $\frac{3}{4}$ for the difference of their inverse squares), the Rydberg formula provides correct values in the special case of K-alpha lines, since the transition in question is the K-alpha transition of the electron from the 1s orbital to the 2p orbital. This is analogous to the Lyman-alpha line transition for hydrogen, and has the same frequency factor. Because the 2p electron is not screened by any other electrons in the atom from the nucleus, the nuclear charge is diminished only by the single remaining 1s electron, causing the system to be effectively a hydrogenic atom, but with a diminished nuclear charge Z−1. Its frequency is thus the Lyman-alpha hydrogen frequency, increased by a factor of $(Z-1)^2$. This formula of $f = c/\lambda$ = (Lyman-alpha frequency)$\cdot(Z-1)^2$ is historically known as Moseley's law (having added a factor ***c*** to convert wavelength to frequency), and can be used to predict wavelengths of the K_α (K-alpha) X-ray spectral emission lines of chemical elements from aluminum to gold. See the biography of Henry Moseley for the historical importance of this law, which was derived empirically at about the same time it was explained by the Bohr model of the atom. For other spectral transitions in multi-electron atoms, the Rydberg formula generally provides incorrect results, since the magnitude of the screening of inner electrons for outer-electron transitions is variable and not possible to compensate for in the simple manner above.

REFERENCES

1. P.W. Atkins (1974). Quanta: A handbook of concepts. Oxford University Press. p. 52. ISBN 0-19-855493-1.
2. T.Hey, P.Walters (2009). The New Quantum Universe. Cambridge University Press. p. 228. ISBN 978-0-521-56457-1.
3. Dirac, P.A.M. (1982). Principles of Quantum Mechanics. International Series of Monographs on Physics (4th ed.). Oxford University Press. p. 255. ISBN 978-0-19-852011-5.
4. see for example Brian Pendleton: Quantum Theory 2012/2013, section 4.3 The Dirac Equation
5. Tommy Ohlsson (22 September 2011). Relativistic Quantum Physics: From Advanced Quantum Mechanics to Introductory Quantum Field Theory. Cambridge University Press. p. 86. ISBN 978-1-139-50432-4.
6. Akhmeteli, Andrey (2011). "One real function instead of the Dirac spinor function" (PDF). Journal of Mathematical Physics. 52 (8): 082303. arXiv:1008.4828 . Bibcode:2011JMP....52h2303A. doi:10.1063/1.3624336.
7. Gisela Dirac-Wahrenburg. "Paul Dirac". Dirac.ch. Retrieved 2013-07-12.
8. Wolfgang Pauli (1927) Zur Quantenmechanik des magnetischen Elektrons Zeitschrift für Physik (43) 601-623
9. Bransden, BH; Joachain, CJ (1983). Physics of Atoms and Molecules (1st ed.). Prentice Hall. p. 638-638. ISBN 0-582-44401-2.

CHAPTER 6

Modern Interpretation of Quantum Mechanics

Chapter Contents

- **Introduction**
- **Klein–Gordon Equation**
- **Pauli Equation**
- **Rydberg Formula**

There are presently several interpretations of quantum mechanics. Many, perhaps most, are attempts to eliminate the element of chanceor indeterminism that is involved in the so-called collapse of the wave function. Information physics claims there is only one world, the quantum world, and the quantum to classical transition» occurs for any large macroscopic object that contains a large number of atoms. For large enough systems, independent quantum events are averaged over. The uncertainty in position and momentum of the object becomes less than the observational uncertainty $\Delta v\ \Delta x \geq h / m$ goes to zero as h / m goes to zero.

INTRODUCTION

In quantum mechanics, the consistent histories (also referred to as decoherent histories) approach is intended to give a modern interpretation of quantum mechanics, generalising the conventional Copenhagen interpretation and providing a natural interpretation of quantum cosmology. This interpretation of quantum mechanics is based on a consistency criterion that then allows probabilities to be assigned to various alternative histories of a system such that the probabilities for each history obey the rules of classical probability while being consistent with the Schrödinger equation. In contrast to some interpretations of quantum mechanics, particularly the Copenhagen interpretation, the framework does not include wavefunction collapse as a relevant description of any physical process, and emphasizes that measurement theory is not a fundamental ingredient of quantum mechanics.

A homogeneous history H_i labels different histories) is a sequence of Propositions $P_{i,j}$ specified at different moments of time $t_{i,j}$ (here j labels the times). We write this as:

$$H_i = (P_{i,1}, P_{i,2}, \ldots, P_{i,n_i})$$

and read it as "the proposition $P_{i,1}$ is true at time $t_{i,1}$ *and then* the proposition $P_{i,2}$ is true at time $t_{i,2}$ *and then* …". The times $t_{i,1} < t_{i,2} < \ldots < t_{i,n_i}$ are strictly ordered and called the *temporal support* of the history.

Inhomogeneous histories are multiple-time propositions which cannot be represented by a homogeneous history. An example is the logical OR of two homogeneous histories: $H_i \vee H_j$.

These propositions can correspond to any set of questions that include all possibilities. Examples might be the three propositions meaning "the electron went through the left slit", "the electron went through the right slit" and "the electron didn't go through either slit". One of the aims of the theory is to show that classical questions such as, "where are my keys?" are consistent. In this case one might use a large number of propositions each one specifying the location of the keys in some small region of space.

Each single-time proposition $P_{i,j}$ can be represented by a projection operator $\hat{P}_{i,j}$ acting on the system's Hilbert space (we use "hats" to denote operators). It is then useful to represent homogeneous histories by the time-ordered tensor product of their single-time projection operators. This is the history projection operator (HPO) formalism developed by Christopher Isham and naturally encodes the logical structure of the history propositions. The homogeneous history H_i is represented by the projection operator

$$\hat{H}_i = \hat{P}_{i,1} \otimes \hat{P}_{i,2} \otimes \cdots \otimes \hat{P}_{i,n_i}$$

This definition can be extended to define projection operators that represent inhomogeneous histories too.

Consistency

An important construction in the consistent histories approach is the **class operator** for a homogeneous history:

$$\hat{C}_{H_i} := T\prod_{j=1}^{n_i} \hat{P}_{i,j}(t_{i,j}) = \hat{P}_{i,1}\hat{P}_{i,2}\cdots\hat{P}_{i,n_i}$$

The symbol T indicates that the factors in the product are ordered chronologically according to their values of $t_{i,j}$: he "past" operators with smaller values of t appear on the right side, and the "future" operators with greater values of t appear on the left side. This definition can be extended to inhomogeneous histories as well.

Central to the consistent histories is the notion of consistency. A set of histories $\{H_i\}$ is **consistent** (or **strongly consistent**) if

$$\mathrm{Tr}(\hat{C}_{H_i}\rho\hat{C}^{\dagger}_{H_j}) = 0$$

for all $i \neq j$. Here ρ represents the initial density matrix, and the operators are expressed in the Heisenberg picture.

The set of histories is weakly consistent if

$$\mathrm{Tr}(\hat{C}_{H_i}\rho\hat{C}^{\dagger}_{H_j}) \approx 0$$

for all $i \neq j$.

Probabilities

If a set of histories is consistent then probabilities can be assigned to them in a consistent way. We postulate that the probability of history H_i is simply

$$\mathrm{Pr}(H_i) = \mathrm{Tr}(\hat{C}_{H_i}\rho\hat{C}^{\dagger}_{H_i})$$

which obeys the axioms of probability if the histories H_i come from the same (strongly) consistent set. As an example, this means the probability of "H_i OR H_j" equals the probability of "H_i" plus the probability of "H_j" minus the probability of "H_i AND H_j", and so forth.

INTERPRETATION

The interpretation based on consistent histories is used in combination with the insights about quantum decoherence. Quantum decoherence implies that irreversible macroscopic phenomena (hence, all classical measurements) render histories automatically consistent, which allows one to recover classical reasoning and "common sense" when applied to the outcomes of these measurements. More precise analysis of decoherence allows (in principle) a quantitative calculation of the boundary between the classical domain and the quantum domain covariance. According to Roland Omnès,

History approach, although it was initially independent of the Copenhagen approach, is in some sense a more elaborate version of it. It has, of course, the advantage of being more precise, of including classical physics, and of providing an explicit logical framework for indisputable proofs. But, when the Copenhagen interpretation is completed by the modern results about correspondence and decoherence, it essentially amounts to the same physics.

Three main differences:

1. The logical equivalence between an empirical datum, which is a macroscopic phenomenon, and the result of a measurement, which is a quantum property, becomes clearer in the new approach, whereas it remained mostly tacit and questionable in the Copenhagen formulation.

2. There are two apparently distinct notions of probability in the new approach. One is abstract and directed toward logic, whereas the other is empirical and expresses the randomness of measurements. We need to understand their relation and why they coincide with the empirical notion entering into the Copenhagen rules.
3. The main difference lies in the meaning of the reduction rule for 'wave packet collapse'. In the new approach, the rule is valid but no specific effect on the measured object can be held responsible for it. Decoherence in the measuring device is enough.

In order to obtain a complete theory, the formal rules above must be supplemented with a particular Hilbert space and rules that govern dynamics, for example a Hamiltonian.

In the opinion of others this still does not make a complete theory as no predictions are possible about which set of consistent histories will actually occur. That is the rules of Consistent Histories, the Hilbert space, and the Hamiltonian must be supplemented by a set selection rule. However, Griffiths holds the opinion that asking the question of which set of histories will "actually occur" is a misinterpretation of the theory; histories are a tool for description of reality, not separate alternate realities.

The proponents of this Consistent Histories interpretation, such as Murray Gell-Mann, James Hartle, Roland Omnès and Robert B. Griffiths argue that their interpretation clarifies the fundamental disadvantages of the old Copenhagen interpretation, and can be used as a complete interpretational framework for quantum mechanics.

In *Quantum Philosophy*, Roland Omnès provides a less mathematical way of understanding this same formalism. The consistent histories approach can be interpreted as a way of understanding which sets of classical questions can be consistently asked of a single quantum system, and which sets of questions are fundamentally inconsistent, and thus meaningless when asked together. It thus becomes possible to demonstrate formally why it is that the questions which Einstein, Podolsky and Rosen assumed could be asked together, of a single quantum system, simply cannot be asked together. On the other hand, it also becomes possible to demonstrate that classical, logical reasoning often does apply, even to quantum experiments – but we can now be mathematically exact about the limits of classical logic.

COPENHAGEN INTERPRETATION

The Copenhagen interpretation is an expression of the meaning of quantum mechanics that was largely devised in the years 1925 to 1927 by Niels Bohr and Werner Heisenberg. It remains one of the most commonly taught interpretations of quantum mechanics.

According to the Copenhagen interpretation, physical systems generally do not have definite properties prior to being measured, and quantum mechanics can only predict the probabilities that measurements will produce certain results. The act of measurement affects the system, causing the set of probabilities to reduce to only one of the possible values immediately after the measurement. This feature is known as wave function collapse.

There have been many objections to the Copenhagen Interpretation over the years. These include: discontinuous jumps when there is an observation, the probabilistic element introduced upon observation, the subjectiveness of requiring an observer, the difficulty of defining a measuring device, and to the necessity of invoking classical physics to describe the laboratory in which the results are measured.

Alternatives to the Copenhagen Interpretation include the many-worlds interpretation, the De Broglie-Bohm (pilot-wave) interpretation, and quantum decoherence theories.

Background

In the early work of Max Planck, Albert Einstein, and Niels Bohr, the occurrence of energy in discrete quantities was postulated in order to explain phenomena such as the spectrum of black-body radiation, the photoelectric effect, and the stability and spectrum of atoms. These phenomena had eluded explanation by classical physics and even appeared to be in contradiction with it. While elementary particles show predictable properties in many experiments, they become thoroughly unpredictable in others, such as attempts to identify individual particle trajectories through a simple physical apparatus.

Classical physics draws a distinction between particles and waves. It also relies on continuity and determinism in natural phenomena. In the early twentieth century, newly discovered atomic and subatomic phenomena seemed to defy those conceptions. In 1925–1926, quantum mechanics was invented as a mathematical formalism that accurately describes the experiments, yet appears to reject those classical conceptions. Instead, it posits that probability, and discontinuity, are fundamental in the physical world. Classical physics also relies on causality. The standing of causality for quantum mechanics is disputed.

Quantum mechanics cannot easily be reconciled with everyday language and observation. Its interpretation has often seemed counter-intuitive to physicists, including its inventors.

The Copenhagen interpretation intends to indicate the proper ways of thinking and speaking about the physical meaning of the mathematical formulations of quantum mechanics and the corresponding experimental results. It offers due respect to discontinuity, probability, and a conception of wave–particle dualism. In some respects, it denies standing to causality.

Origin of the Term

Werner Heisenberg had been an assistant to Niels Bohr at his institute in Copenhagen during part of the 1920s, when they helped originate quantum mechanical theory. In 1929, Heisenberg gave a series of invited lectures at the University of Chicago explaining the new field of quantum mechanics. The lectures then served as the basis for his textbook, *The Physical Principles of the Quantum Theory*, published in 1930. In the book's preface, Heisenberg wrote:

On the whole the book contains nothing that is not to be found in previous publications, particularly in the investigations of Bohr. The purpose of the book seems to me to be fulfilled if it contributes somewhat to the diffusion of that 'Kopenhagener Geist der Quantentheorie' if I may so express myself, which has directed the entire development of modern atomic physics. The term 'Copenhagen interpretation' suggests something more than just a spirit, such as some definite set of rules for interpreting the mathematical formalism of quantum mechanics, presumably dating back to the 1920s. However, no such text exists, apart from some informal popular lectures by Bohr and Heisenberg, which contradict each other on several important issues. It appears that the particular term, with its more definite sense, was coined by Heisenberg in the 1950s, while criticizing alternate "interpretations" (e.g., David Bohm's) that had been developed. Lectures with the titles 'The Copenhagen Interpretation of Quantum Theory' and 'Criticisms and Counterproposals to the Copenhagen Interpretation', that Heisenberg delivered in 1955, are reprinted in the collection *Physics and Philosophy*. Before the book was released for sale, Heisenberg privately expressed regret for having used the term, due to its suggestion of the existence of other interpretations, that he considered to be "nonsense".

Current Status of the Term

According to an opponent of the Copenhagen interpretation, John G. Cramer, "Despite an extensive literature which refers to, discusses, and criticizes the Copenhagen interpretation of quantum mechanics, nowhere does there seem to be any concise statement which defines the full Copenhagen interpretation."

Principles

Because it consists of the views developed by a number of scientists and philosophers during the second quarter of the 20th Century, there is no uniquely definitive statement of the Copenhagen interpretation. Moreover, by different commentators and researchers, various ideas have been associated with it; Asher Peres remarked that very different, sometimes opposite, views are presented as "the Copenhagen interpretation" by different authors. Nonetheless, there are several basic principles that are generally accepted as being part of the interpretation:

- A wave function {\displaystyle \Psi } represents the state of the system. It encapsulates everything that can be known about that system before an observation; there are no additional "hidden parameters". The wavefunction evolves smoothly in time while isolated from other systems.
- The properties of the system are subject to a principle of incompatibility. Certain properties cannot be jointly defined for the same system at the same time. The incompatibility is expressed quantitatively by Heisenberg›s uncertainty principle. For example, if a particle at a particular instant has a definite location, it is meaningless to speak of its momentum at that instant.
- During an observation, the system must interact with a laboratory device. When that device makes a measurement, the wave function of the systems is said to collapse, or irreversibly reduce to an eigenstate of the observable that is registered.
- The results provided by measuring devices are essentially classical, and should be described in ordinary language. This was particularly emphasized by Bohr,

and was accepted by Heisenberg.

- The description given by the wave function is probabilistic. This principle is called the Born rule, after Max Born.
- The wave function expresses a necessary and fundamental wave–particle duality. This should be reflected in ordinary language accounts of experiments. An experiment can show particle-like properties, or wave-like properties, according to the complementarity principle of Niels Bohr.
- The inner workings of atomic and subatomic processes are necessarily and essentially inaccessible to direct observation, because the act of observing them would greatly affect them.
- When quantum numbers are large, they refer to properties which closely match those of the classical description. This is the correspondence principle of Bohr and Heisenberg.

Metaphysics of the Wave Function

The Copenhagen Interpretation denies that the wave function provides a directly apprehensible image of an ordinary material body or a discernible component of some such, or anything more than a theoretical concept.

In metaphysical terms, the Copenhagen interpretation views quantum mechanics as providing knowledge of phenomena, but not as pointing to 'really existing objects', which it regarded as residues of ordinary intuition. This makes it an epistemic theory. This may be contrasted with Einstein's view, that physics should look for 'really existing objects', making itself an ontic theory.

The metaphysical question is sometimes asked: "Could quantum mechanics be extended by adding so-called "hidden variables" to the mathematical formalism, to convert it from an epistemic to an ontic theory?" The Copenhagen interpretation answers this with a strong 'No'. It is sometimes alleged, for example by J.S. Bell, that Einstein opposed the Copenhagen interpretation because he believed that the answer to that question of "hidden variables" was 'Yes'. That allegation has achieved mythical potency, but is mistaken. Countering that myth, Max Jammer writes "Einstein never proposed a hidden variable theory." Einstein explored the possibility of a hidden variable theory, and wrote a paper describing his exploration, but withdrew it from publication because he felt it was faulty.

Because it asserts that a wave function becomes 'real' only when the system is observed, the term "subjective" is sometimes proposed for the Copenhagen interpretation. This term is rejected by many Copenhagenists because the process of observation is mechanical and does not depend on the individuality of the observer.

Some authors have proposed that Bohr was influenced by positivism (or even pragmatism). On the other hand, Bohr and Heisenberg were not in complete agreement, and they held different views at different times. Heisenberg in particular was prompted to move towards realism.

Carl Friedrich von Weizsäcker, while participating in a colloquium at Cambridge, denied that the Copenhagen interpretation asserted "What cannot be observed does not exist." He suggested instead that the Copenhagen interpretation follows the principle "What is observed certainly exists; about what is not observed we are still free to make suitable assumptions. We use that freedom to avoid paradoxes."

Born Rule

Max Born speaks of his probability interpretation as a "statistical interpretation" of the wave function, and the Born rule is essential to the Copenhagen interpretation. But writers do not all follow the same terminology.

The phrase 'statistical interpretation', referring to the 'ensemble interpretation', often indicates an interpretation of the Born rule somewhat different from the Copenhagen interpretation. For the Copenhagen interpretation, it is axiomatic that the wave function exhausts all that can ever be known in advance about a particular occurrence of the system. The 'statistical' or 'ensemble' interpretation, on the other hand, is explicitly agnostic about whether the information in the wave function is exhaustive of what might be known in advance. It sees itself as more 'minimal' than the Copenhagen interpretation in its claims. It only goes as far as saying that on every occasion of observation, some actual value of some property is found, and that such values are found probabilistically, as detected by many occasions of observation of the same system. The many occurrences of the system are said to constitute an 'ensemble', and they jointly reveal the probability through these occasions of observation. Though they all have the same wave function, the elements of the ensemble might not be identical to one another in all respects, according to the 'agnostic' interpretations. They may, for all we know, beyond current knowledge and beyond the wave function, have individual distinguishing properties. For present-day science, the experimental significance of these various forms of Born's rule is the same, since they make the same predictions about the probability distribution of outcomes of observations, and the unobserved or unactualized potential properties are not accessible to experiment.

Nature of Collapse

Those who hold to the Copenhagen interpretation are willing to say that a wave function involves the various probabilities that a given event will proceed to certain different outcomes. But when the apparatus registers one of those outcomes, no probabilities or superposition of the others linger.

According to Howard, wave function collapse is not mentioned in the writings of Bohr.

Some argue that the concept of the collapse of a "real" wave function was introduced by Heisenberg and later developed by John von Neumann in 1932. However, Heisenberg spoke of the wavefunction as representing available knowledge of a system, and did not use the term "collapse" per se, but instead termed it "reduction" of the wavefunction to a new state representing the change in available knowledge which occurs once a particular phenomenon is registered by the apparatus (often called "measurement").

In 1952 David Bohm developed decoherence, an *explanatory mechanism* for the *appearance* of wave function collapse. Bohm applied decoherence to Louis DeBroglie›s pilot wave theory, producing Bohmian mechanics, the first successful hidden variables interpretation of quantum mechanics. Collapse was avoided by Hugh Everett in 1957 in his relative state interpretation. Decoherence was largely ignored until the 1980s.

Non-Separability of the Wave Function

The domain of the wave function is configuration space, an abstract object quite different from ordinary physical space–time. At a single "point" of configuration space, the wave function collects probabilistic information about several distinct particles, that respectively have physically space-like separation. So the wave function is said to supply a non-separable representation. This reflects a feature of the quantum world that was recognized by Einstein as early as 1905.

In 1927, Bohr drew attention to a consequence of non-separability. The evolution of the system, as determined by the Schrödinger equation, does not display particle trajectories through space–time. It is possible to extract trajectory information from such evolution, but not simultaneously to extract energy–momentum information. This incompatibility is expressed in the Heisenberg uncertainty principle. The two kinds of information have to be extracted on different occasions, because of the non-separability of the wave function representation. In Bohr's thinking, space–time visualizability meant trajectory information. Again, in Bohr's thinking, 'causality' referred to energy–momentum transfer; in his view, lack of energy–momentum knowledge meant lack of 'causality' knowledge. Therefore Bohr thought that knowledge respectively of 'causality' and of space–time visualizability were incompatible but complementary.

Wave–Particle Dilemma

The term Copenhagen interpretation is not well defined when one asks about the wave–particle dilemma, because Bohr and Heisenberg had different or perhaps disagreeing views on it.

According to Camilleri, Bohr thought that the distinction between a wave view and a particle view was defined by a distinction between experimental setups, while, differing, Heisenberg thought that it was defined by the possibility of viewing the mathematical formulas as referring to waves or particles. Bohr thought that a particular experimental setup would display either a wave picture or a particle picture, but not both. Heisenberg thought that every mathematical formulation was capable of both wave and particle interpretations.

Alfred Landé was for a long time considered orthodox. He did, however, take the Heisenberg viewpoint, in so far as he thought that the wave function was always mathematically open to both interpretations. Eventually this led to his being considered unorthodox, partly because he did not accept Bohr's one-or-the-other view, preferring Heisenberg's always-both view. Another part of the reason for branding Landé unorthodox was that he recited, as did Heisenberg, the 1923 work of old-quantum-theorist William Duane, which anticipated a quantum mechanical theorem that had not been recognized by Born. That theorem seems to make the always-both view, like the one adopted by Heisenberg, rather cogent. One might say It›s there in the mathematics, but that is not a physical statement that would have convinced Bohr. Perhaps the main reason for attacking Landé is that his work demystified the phenomenon of diffraction of particles of matter, such as buckyballs.

Acceptance Among Physicists

Throughout much of the twentieth century the Copenhagen interpretation had overwhelming acceptance among physicists. Although astrophysicist and science writer John Gribbin described it as having fallen from primacy after the 1980s, according to a poll conducted at a quantum mechanics conference in 1997, the Copenhagen interpretation remained the most widely accepted specific interpretation of quantum mechanics among physicists. In more recent polls conducted at various quantum mechanics conferences, varying results have been found.

Consequences

The nature of the Copenhagen Interpretation is exposed by considering a number of experiments and paradoxes.

1. Schrödinger's cat

This thought experiment highlights the implications that accepting uncertainty at the microscopic level has on macroscopic objects. A cat is put in a sealed box, with its life or death made dependent on the state of a subatomic particle. Thus a description of the cat during the course of the experiment—having been entangled with the state of a subatomic particle—becomes a "blur" of "living and dead cat." But this can't be accurate because it implies the cat is actually both dead and alive until the box is opened to check on it. But the cat, if it survives, will only remember being alive. Schrödinger resists "so naively accepting as valid a 'blurred model' for representing reality." *How can the cat be both alive and dead?*

The Copenhagen Interpretation: The wave function reflects our knowledge of the system. The wave function $(|\text{dead}\rangle + |\text{alive}\rangle)/\sqrt{2}$ means that, once the cat is observed, there is a 50% chance it will be dead, and 50% chance it will be alive.

2. Wigner's Friend

Wigner puts his friend in with the cat. The external observer believes the system is in the state $(|\text{dead}\rangle + |\text{alive}\rangle)/\sqrt{2}$. His friend, however, is convinced that the cat is alive, i.e. for him, the cat is in the state $|\text{alive}\rangle$

How can Wigner and his friend see different wave functions?

The Copenhagen Interpretation: The answer depends on the positioning of Heisenberg cut, which can be placed arbitrarily. If Wigner's friend is positioned on the same side of the cut as the external observer, his measurements collapse the wave function for both observers. If he is positioned on the cat's side, his interaction with the cat is not considered a measurement.

3. Double-slit diffraction

Light passes through double slits and onto a screen resulting in a diffraction pattern. *Is light a particle or a wave?*

The Copenhagen Interpretation: Light is neither. A particular experiment can demonstrate particle (photon) or wave properties, but not both at the same time (Bohr's Complementarity Principle).

The same experiment can in theory be performed with *any* physical system: electrons, protons, atoms, molecules, viruses, bacteria, cats, humans, elephants, planets, etc. In practice it has been performed for light, electrons, buckminsterfullerene, and some atoms. Due to the smallness of Planck's constant it is practically impossible to realize experiments that directly reveal the wave nature of any system bigger than a few atoms but, in general, quantum mechanics considers all matter as possessing both particle and wave behaviors. The greater systems (like viruses, bacteria, cats, etc.) are considered as "classical" ones but only as an approximation, not exact.

4. EPR (Einstein–Podolsky–Rosen) paradox

Entangled particles are emitted in a single event. Conservation laws ensure that the measured spin of one particle must be the opposite of the measured spin of the other, so that if the spin of one particle is measured, the spin of the other particle is now instantaneously known. The most discomforting aspect of this paradox is that the effect is instantaneous so that something that happens in one galaxy could cause an instantaneous change in another galaxy. But, according to Einstein›s theory of special relativity, no information-bearing signal or entity can travel at or faster than the speed of light, which is finite. Thus, it seems as if the Copenhagen interpretation is inconsistent with special relativity.

The Copenhagen Interpretation: Assuming wave functions are not real, wave-function collapse is interpreted subjectively. The moment one observer measures the spin of one particle, he knows the spin of the other. However, another observer cannot benefit until the results of that measurement have been relayed to him, at less than or equal to the speed of light.

Copenhagenists claim that interpretations of quantum mechanics where the wave function is regarded as real have problems with EPR-type effects, since they imply that the laws of physics allow for influences to propagate at speeds greater than the speed of light. However, proponents of many worlds and the transactional interpretation (TI) maintain that Copenhagen interpretation is fatally non-local.

The claim that EPR effects violate the principle that information cannot travel faster than the speed of light have been countered by noting that they cannot be used for signaling because neither observer can control, or predetermine, what he observes, and therefore cannot manipulate what the other observer measures. However, this is a somewhat spurious argument, in that the speed of light limitation applies to all information, not to what can or cannot be subsequently done with the information. On the other hand, the special theory of relativity contains no notion of information at all. The fact that no classical body can exceed the speed of light (no matter how much acceleration is applied) is a consequence of classical relativistic mechanics. As the correlation between the two particles in an EPR experiment is most probably not established by classical bodies or light signals, the displayed non-locality is not at odds with special relativity.

A further argument against Copenhagen interpretation is that relativistic difficulties about establishing which measurement occurred first or last, or whether they occurred quite at the same time, also undermine the idea that in different instants and measurements different outcomes can occur. The spin would be kept as a constant for a continuous interval of time, i.e. as a real *variable*, and thus it would seem to violate the general rule (of the classic Copenhagen interpretation) that every measurement gives nothing else than a random outcome subject to certain probabilities.

Criticism

The completeness of quantum mechanics (thesis 1) was attacked by the Einstein–Podolsky–Rosen thought experiment which was intended to show that quantum mechanics could not be a complete theory.

Experimental tests of Bell's inequality using particles have supported the quantum mechanical prediction of entanglement.

The Copenhagen Interpretation gives special status to measurement processes without clearly defining them or explaining their peculiar effects. In his article entitled "Criticism and Counterproposals to the Copenhagen Interpretation of Quantum Theory," countering the view of Alexandrov that (in Heisenberg's paraphrase) "the wave function in configuration space characterizes the objective state of the electron." Heisenberg says,

Of course the introduction of the observer must not be misunderstood to imply that some kind of subjective features are to be brought into the description of nature. The observer has, rather, only the function of registering decisions, i.e., processes in space and time, and it does not matter whether the observer is an apparatus or a human being; but the registration, i.e., the transition from the "possible" to the "actual," is absolutely necessary here and cannot be omitted from the interpretation of quantum theory.

Many physicists and philosophers have objected to the Copenhagen interpretation, both on the grounds that it is non-deterministic and that it includes an undefined measurement process that converts probability functions into non-probabilistic measurements. Einstein's comments "I, at any rate, am convinced that He (God) does not throw dice." and "Do you really think the moon isn't there if you aren't looking at it?" exemplify this. Bohr, in response, said, "Einstein, don't tell God what to do."

Steven Weinberg in "Einstein's Mistakes", *Physics Today*, November 2005, page 31, said:

All this familiar story is true, but it leaves out an irony. Bohr's version of quantum mechanics was deeply flawed, but not for the reason Einstein thought. The Copenhagen interpretation describes what happens when an observer makes a measurement, but the observer and the act of measurement are themselves treated classically. This is surely wrong: Physicists and their apparatus must be governed by the same quantum mechanical rules that govern everything else in the universe. But these rules are expressed in terms of a wave function (or, more precisely, a state vector) that evolves in a perfectly deterministic way. So where do the probabilistic rules of the Copenhagen interpretation come from?

Considerable progress has been made in recent years toward the resolution of the problem, which I cannot go into here. It is enough to say that neither Bohr nor Einstein had focused on the real problem with quantum mechanics. The Copenhagen rules clearly work, so they have to be accepted. But this leaves the task of explaining them by applying the deterministic equation for the evolution of the wave function, the Schrödinger equation, to observers and their apparatus.

The problem of thinking in terms of classical measurements of a quantum system becomes particularly acute in the field of quantum cosmology, where the quantum system is the universe.

E. T. Jaynes, from a Bayesian point of view, argued that probability is a measure of a state of information about the physical world. Quantum mechanics under the Copenhagen Interpretation interpreted probability as a physical phenomenon, which is what Jaynes called a Mind Projection Fallacy.

Common criticisms of the Copenhagen interpretation often lead to the problem of continuum of random occurrences: whether in time (as subsequent measurements, which under certain interpretations of the measurement problem may happen continuously) or even in space. A recent experiment showed that a particle may leave a trace about the path which it used when travelling as a wave – and that this trace exhibits equality of both paths. If such result is raised to the rank of a wave-only non-transactional worldview and proved better – i.e. that a particle is in fact a continuum of points capable of acting independently but under a common wavefunction – it would rather support theories such as Bohm's one (with its guiding towards the centre of orbital and spreading of physical properties over it) than interpretations which presuppose full randomness, because with the latter it will be problematic to demonstrate universally and in all practical cases how can a particle remain coherent in time, in spite of non-zero probabilities of its individual points going into regions distant from the centre of mass (through a continuum of different random determinations). An alternative possibility would be to assume that there is a finite number of instants/points within a given time or area, but theories which try to quantize the space or time itself seem to be fatally incompatible with the special relativity.

The view that particle diffraction logically guarantees the need for a wave interpretation has been questioned. A recent experiment has carried out the two-slit protocol with helium atoms. The basic physics of quantal momentum transfer considered here was originally pointed out in 1923, by William Duane, before quantum mechanics was invented. It was later recognized by Heisenberg and by Pauling. It was championed against orthodox ridicule by Alfred Landé. It has also recently been considered by Van Vliet. If the diffracting slits are considered as classical objects, theoretically ideally seamless, then a wave interpretation seems necessary, but if the diffracting slits are considered physically, as quantal objects exhibiting collective quantal motions, then the particle-only and wave-only interpretations seem perhaps equally valid.

Alternatives

The Ensemble interpretation is similar; it offers an interpretation of the wave function, but not for single particles. The consistent histories interpretation advertises itself as "Copenhagen done right". Although the Copenhagen interpretation is often confused with the idea that consciousness causes collapse, it defines an "observer" merely as that which collapses the wave function. Quantum information theories are more recent, and have attracted growing support.

Under realism and indeterminism, if the wave function is regarded as ontologically real, and collapse is entirely rejected, a many worlds theory results. If wave function collapse is regarded as ontologically real as well, an objective collapse theory is obtained. Under realism and determinism (as well as non-localism), a hidden variable theory exists, e.g., the de Broglie–Bohm interpretation, which treats the wavefunction as real, position and momentum as definite and resulting from the expected values, and physical properties as spread in space. For an atemporal indeterministic interpretation that "makes no attempt to give a 'local' account on the level of determinate particles", the conjugate wavefunction, ("advanced" or time-reversed) of the relativistic version of the wavefunction, and the so-called "retarded" or time-forward version are both regarded as real and the transactional interpretation results.

Many physicists have subscribed to the instrumentalist interpretation of quantum mechanics, a position often equated with eschewing all interpretation. It is summarized by the sentence "Shut up and calculate!". While this slogan is sometimes attributed to Paul Dirac or Richard Feynman, it seems to be due to David Mermin.

DE BROGLIE–BOHM THEORY

The de Broglie–Bohm theory, also known as the pilot-wave theory, Bohmian mechanics, the Bohm or Bohm's interpretation, and the causal interpretation, is an interpretation of quantum theory. In addition to a wavefunction on the space of all possible configurations, it also postulates an actual configuration that exists even when unobserved. The evolution over time of the configuration (that is, the positions of all particles or the configuration of all fields) is defined by the wave function by a guiding equation. The evolution of the wave function over time is given by Schrödinger›s equation. The theory is named after Louis de Broglie (1892–1987) and David Bohm (1917–1992).

The theory is deterministic and explicitly nonlocal: the velocity of any one particle depends on the value of the guiding equation, which depends on the configuration of the system given by its wavefunction; the latter depends on the boundary conditions of the system, which in principle may be the entire universe.

The theory results in a measurement formalism, analogous to thermodynamics for classical mechanics, that yields the standard quantum formalism generally associated with the Copenhagen interpretation. The theory›s explicit non-locality resolves the measurement problem, which is conventionally delegated to the topic of interpretations of quantum mechanics in the Copenhagen interpretation. The Born rule in Broglie–Bohm theory is not a basic law. Rather, in this theory the link between the probability density and the wave function has the status of a hypothesis, called the quantum equilibrium hypothesis, which is additional to the basic principles governing the wave function.

The theory was historically developed in the 1920s by de Broglie, who in 1927 was persuaded to abandon it in favour of the then-mainstream Copenhagen interpretation. David Bohm, dissatisfied with the prevailing orthodoxy, rediscovered de Broglie›s pilot-wave theory in 1952. Bohm›s suggestions were not widely received then, partly due to reasons unrelated to their content, connected to Bohm›s youthful communist affiliations. De Broglie–Bohm theory was widely deemed unacceptable by mainstream theorists, mostly because of its explicit non-locality. Bell›s theorem (1964) was inspired by Bell›s discovery of the work of David Bohm and his subsequent wondering whether the obvious nonlocality of the theory could be eliminated. Since the 1990s, there has been renewed interest in formulating extensions to de Broglie–Bohm theory, attempting to reconcile it with special relativity and quantum field theory, besides other features such as spin or curved spatial geometries.

The *Stanford Encyclopedia of Philosophy* article on Quantum decoherence (Guido Bacciagaluppi, 2012) groups approaches to quantum mechanics into five groups, of which pilot-wave theories are one (the others being the Copenhagen interpretation, objective collapse theories, many-world interpretations and modal interpretations).

There are several equivalent mathematical formulations of the theory, and it is known by a number of different names. The de Broglie wave has a macroscopic analogy termed Faraday wave.

De Broglie–Bohm theory is based on the following postulates:

There is a configuration q of the universe, described by coordinates q^k, which is an element of the configuration space Q The configuration space is different for different versions of pilot-wave theory. For example, this may be the space of positions $\mathbf{Q}_k$ of N particles, or, in case of field theory, the space of field configurations $\phi(x)$. he configuration evolves (for spin=0) according to the guiding equation

$$m_k \frac{dq^k}{dt}(t) = \hbar \nabla_k \operatorname{Im} \ln \psi(q,t) = \hbar \operatorname{Im}\left(\frac{\nabla_k \psi}{\psi}\right)(q,t) = \frac{m_k \mathbf{j}_k}{\psi^* \psi} = \operatorname{Re}\left(\frac{\hat{\mathbf{P}}_k \Psi}{\Psi}\right)$$

Where J is the probability current or probability flux, and $\hat{\mathbf{P}}$ is the momentum operator. Here, $\psi(q,t)$ is the standard complex-valued wavefunction known from quantum theory, which evolves according to Schrödinger's equation

$$i\hbar \frac{\partial}{\partial t}\psi(q,t) = -\sum_{i=1}^{N} \frac{\hbar^2}{2m_i} \nabla_i^2 \psi(q,t) + V(q)\psi(q,t).$$

This already completes the specification of the theory for any quantum theory with Hamilton operator of type

$$H = \sum \frac{1}{2m_i} \hat{p}_i^2 + V(\hat{q}).$$

This already completes the specification of the theory for any quantum theory with Hamilton operator of type $H = \sum \frac{1}{2m_i} \hat{p}_i^2 + V(\hat{q})$

This already completes the specification of the theory for any quantum theory with Hamilton operator of type $H = \sum \frac{1}{2m_i}\hat{p}_i^2 + V(\hat{q})$

The configuration is distributed according to $|\psi(q,t)|^2$ at some moment of time t

- and this consequently holds for all times. Such a state is named quantum equilibrium. With quantum equilibrium, this theory agrees with the results of standard quantum mechanics.

Notably, even though this latter relation is frequently presented as an axiom of the theory, in Bohm's original papers of 1952 it was presented as derivable from statistical-mechanical arguments. This argument was further supported by the work of Bohm in 1953 and was substantiated by Vigier and Bohm's paper of 1954, in which they introduced stochastic *fluid fluctuations* that drive a process of asymptotic relaxation from quantum non-equilibrium to quantum equilibrium ($\rho \rightarrow |\psi|^2$).

Double-Slit Experiment

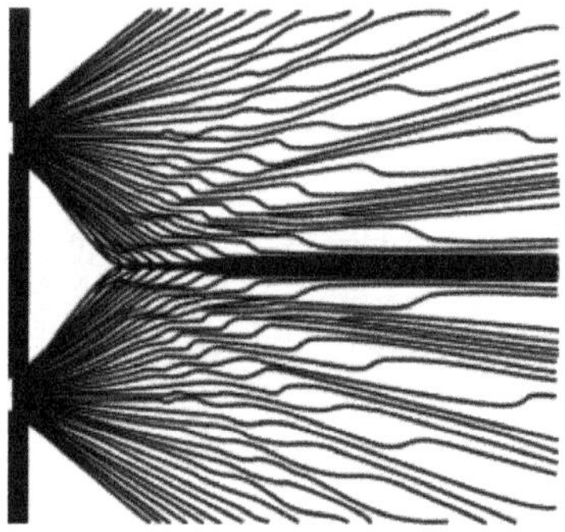

The Bohmian trajectories for an electron going through the two-slit experiment. A similar pattern was also extrapolated from weak measurements of single photons.

The double-slit experiment is an illustration of wave-particle duality. In it, a beam of particles (such as electrons) travels through a barrier that has two slits. If one puts a detector screen on the side beyond the barrier, the pattern of detected particles shows interference fringes characteristic of waves arriving at the screen from two sources (the two slits); however, the interference pattern is made up of individual dots corresponding to particles that had arrived on the screen. The system seems to exhibit the behaviour of both waves (interference patterns) and particles (dots on the screen).

If we modify this experiment so that one slit is closed, no interference pattern is observed. Thus, the state of both slits affects the final results. We can also arrange to have a minimally invasive detector at one of the slits to detect which slit the particle went through. When we do that, the interference pattern disappears.

The Copenhagen interpretation states that the particles are not localised in space until they are detected, so that, if there is no detector on the slits, there is no information about which slit the particle has passed through. If one slit has a detector on it, then the wavefunction collapses due to that detection. In de Broglie–Bohm theory, the wavefunction is defined at both slits, but each particle has a well-defined trajectory that passes through exactly one of the slits. The final position of the particle on the detector screen and the slit through which the particle passes is determined by the initial position of the particle. Such initial position is not knowable or controllable by the experimenter, so there is an appearance of randomness in the pattern of detection. In Bohm's 1952 papers he used the wavefunction to construct a quantum potential that, when included in Newton's equations, gave the trajectories of the particles streaming through the two slits. In effect the wavefunction interferes with itself and guides the particles by the quantum potential in such a way that the particles avoid the regions in which the interference is destructive and are attracted to the regions in which the interference is constructive, resulting in the interference pattern on the detector screen.

To explain the behavior when the particle is detected to go through one slit, one needs to appreciate the role of the conditional wavefunction and how it results in the collapse of the wavefunction; this is explained below. The basic idea is that the environment registering the detection effectively separates the two wave packets in configuration space.

The Theory

The Ontology

The ontology of de Broglie–Bohm theory consists of a configuration $q(t) \in Q$ of the universe and a pilot wave $\psi(q,t) \in \mathbb{C}$. The configuration space Q can be chosen differently, as in classical mechanics and standard quantum mechanics.

Thus, the ontology of pilot-wave theory contains as the trajectory $q(t) \in Q$ we know from classical mechanics, as the wavefunction $\psi(q,t) \in \mathbb{C}$ of quantum theory. So, at every moment of time there exists not only a wavefunction, but also a well-defined configuration of the whole universe (i.e., the system as defined by the boundary conditions used in solving the Schrödinger equation). The correspondence to our experiences is made by the identification of the configuration of our brain with some part of the configuration of the whole universe $q(t) \in Q$, as in classical mechanics.

While the ontology of classical mechanics is part of the ontology of de Broglie–Bohm theory, the dynamics are very different. In classical mechanics, the accelerations of the particles are imparted directly by forces, which exist in physical three-dimensional space. In de Broglie–Bohm theory, the velocities of the particles are given by the wavefunction, which exists in a $3N$-dimensional configuration space, where N corresponds to the number of particles in the system; Bohm hypothesized that each particle has a "complex and subtle inner structure" that provides the capacity to react to the information provided by the wavefunction by the quantum potential. Also, unlike in classical mechanics, physical properties (e.g., mass, charge) are spread out over the wavefunction in de Broglie–Bohm theory, not localized at the position of the particle.

The wavefunction itself, and not the particles, determines the dynamical evolution of the system: the particles do not act back onto the wave function. As Bohm and Hiley worded it, the Schrödinger equation for the quantum field does not have sources, nor does it have any other way by which the field could be directly affected by the condition of the particles the quantum theory can be understood completely in terms of the assumption that the quantum field has no sources or other forms of dependence on the particles. P. Holland considers this lack of reciprocal action of particles and wave function to be one mong the many nonclassical properties exhibited by this theory. It should be noted, however, that Holland has later called this a merely *apparent* lack of back reaction, due to the incompleteness of the description.

In what follows below, we will give the setup for one particle moving in $\mathbb{R}^3$ followed by the setup for *N* particles moving in 3 dimensions. In the first instance, configuration space and real space are the same, while in the second, real space is still $\mathbb{R}^3$ but configuration space becomes $\mathbb{R}^{3N}$ While the particle positions themselves are in real space, the velocity field and wavefunction are on configuration space, which is how particles are entangled with each other in this theory.

Extensions to this theory include spin and more complicated configuration spaces. We use variations of Q for particle positions, while ψ represents the complex-valued wavefunction on configuration space.

Guiding Equation

For a spinless single particle moving in $\mathbb{R}^3$, the particle's velocity is given by

$$\frac{d\mathbf{Q}}{dt}(t) = \frac{\hbar}{m}\operatorname{Im}\left(\frac{\nabla\psi}{\psi}\right)(\mathbf{Q},t).$$

For many particles, we label them as $\mathbf{Q}_k$ for the k-th particle, and their velocities are given by

$$\frac{d\mathbf{Q}_k}{dt}(t) = \frac{\hbar}{m_k}\operatorname{Im}\left(\frac{\nabla_k\psi}{\psi}\right)(\mathbf{Q}_1,\mathbf{Q}_2,\ldots,\mathbf{Q}_N,t).$$

The main fact to notice is that this velocity field depends on the actual positions of all of the N particles in the universe. As explained below, in most experimental situations, the influence of all of those particles can be encapsulated into an effective wavefunction for a subsystem of the universe.

Schrödinger's Equation

The one-particle Schrödinger equation governs the time evolution of a complex-valued wavefunction on $\mathbb{R}^3$. The equation represents a quantized version of the total energy of a classical system evolving under a real-valued potential function V on $\mathbb{R}^3$.

$$i\hbar\frac{\partial}{\partial t}\psi = -\frac{\hbar^2}{2m}\nabla^2\psi + V\psi.$$

For many particles, the equation is the same except that ψ and V are now on configuration space, $\mathbb{R}^{3N}$:

$$i\hbar\frac{\partial}{\partial t}\psi = -\sum_{k=1}^{N}\frac{\hbar^2}{2m_k}\nabla_k^2\psi + V\psi.$$

This is the same wavefunction as in conventional quantum mechanics.

Relation to the Born Rule

In Bohm's original papers, he discusses how de Broglie–Bohm theory results in the usual measurement results of quantum mechanics. The main idea is that this is true if the positions of the particles satisfy the statistical distribution given by $|\psi|^2$ And that distribution is guaranteed to be true for all time by the guiding equation if the initial distribution of the particles satisfies $|\psi|^2$

For a given experiment, we can postulate this as being true and verify experimentally that it does indeed hold true, as it does. But, as argued in Dürr et al., one needs to argue that this distribution for subsystems is typical. They argue that $|\psi|^2$ by virtue of its equivariance under the dynamical evolution of the system, is the appropriate measure of typicality for initial conditions of the positions of the particles. They then prove that the vast majority of possible initial configurations will give rise to statistics obeying the Born rule (i.e., $|\psi|^2$ for measurement outcomes. In summary, in a universe governed by the de Broglie–Bohm dynamics, Born rule behavior is typical.

The situation is thus analogous to the situation in classical statistical physics. A low-entropy initial condition will, with overwhelmingly high probability, evolve into a higher-entropy state: behavior consistent with the second law of thermodynamics is typical. There are, of course, anomalous initial conditions that would give rise to violations of the second law. However, in the absence of some very detailed evidence supporting the actual realization of one of those special initial conditions, it would be quite unreasonable to expect anything but the actually observed uniform increase of entropy. Similarly, in the de Broglie–Bohm theory, there are anomalous initial conditions that would produce measurement statistics in violation of the Born rule (i.e., in conflict with the predictions of standard quantum theory). But the typicality theorem shows that, in the absence of some specific reason to believe that one of those special initial conditions was in fact realized, the Born rule behavior is what one should expect.

It is in that qualified sense that the Born rule is, for the de Broglie–Bohm theory, a theorem rather than (as in ordinary quantum theory) an additional postulate.

It can also be shown that a distribution of particles that is *not* distributed according to the Born rule (that is, a distribution "out of quantum equilibrium") and evolving under the de Broglie–Bohm dynamics is overwhelmingly likely to evolve dynamically into a state distributed as $|\psi|^2$

The Conditional Wavefunction of a Subsystem

In the formulation of the de Broglie–Bohm theory, there is only a wavefunction for the entire universe (which always evolves by the Schrödinger equation). It should, however, be noted that the "universe" is simply the system limited by the same boundary conditions used to solve the Schrödinger equation. However, once the theory is formulated, it is convenient to introduce a notion of wavefunction also for subsystems of the universe. Let us write the wavefunction of the universe as $\psi(t, q^{\mathrm{I}}, q^{\mathrm{II}})$, where q^{I} denotes the configuration variables associated to some subsystem (I) of the universe, and q II denotes the remaining configuration variables. Denote respectively by QI(t) and QII(t) the actual configuration of subsystem (I) and of the rest of the universe. For simplicity, we consider here only the spinless case. The *conditional wavefunction* of subsystem (I) is defined by

$$\psi^{\mathrm{I}}(t, q^{\mathrm{I}}) = \psi(t, q^{\mathrm{I}}, Q^{\mathrm{II}}(t)).$$

It follows immediately from the fact that Q(t) = (QI(t), QII(t)) satisfies a guiding equation identical to the one presented in the formulation of the theory, with the universal wavefunction ψ^{I} eplaced with the conditional wavefunction ψ^{I}. Also, the fact that $Q(t)$ is random with probability density given by the square modulus of $\psi(t, \cdot)$ ψ^{I} Also, the fact that Q(t) is random with probability density given by the square modulus of $\psi(t, \cdot)$ implies that the conditional probability density of $Q^{\mathrm{I}}(t)$ given $Q^{\mathrm{II}}(t)$ is given by the square modulus of the (normalized) conditional wavefunction $\psi^{\mathrm{I}}(t, \cdot)$ (in the terminology of Dürr et al. this fact is called the *fundamental conditional probability formula*).

Unlike the universal wavefunction, the conditional wavefunction of a subsystem does not always evolve by the Schrödinger equation, but in many situations it does. For instance, if the universal wavefunction factors as

$$\psi(t, q^{\mathrm{I}}, q^{\mathrm{II}}) = \psi^{\mathrm{I}}(t, q^{\mathrm{I}})\psi^{\mathrm{II}}(t, q^{\mathrm{II}}) + \phi(t, q^{\mathrm{I}}, q^{\mathrm{II}}),$$

where ϕ solves Schrödinger equation and, $\phi(t, q^{\mathrm{I}}, Q^{\mathrm{II}}(t)) = 0$ for all t and q^{I}. Then, again, the conditional wavefunction of subsystem (I) is (up to an irrelevant scalar factor) equal to ψ^{I}, and if the Hamiltonian does not contain an interaction term between subsystems (I) and (II), then ψ^{I} satisfies a Schrödinger equation.

The fact that the conditional wavefunction of a subsystem does not always evolve by the Schrödinger equation is related to the fact that the usual collapse rule of standard quantum theory emerges from the Bohmian formalism when one considers conditional wavefunctions of subsystems.

Extensions

Relativity

Pilot-wave theory is explicitly nonlocal, which is in ostensible conflict with special relativity. Various extensions of "Bohm-like" mechanics exist that attempt to resolve this problem. Bohm himself in 1953 presented an extension of the theory satisfying the Dirac equation for a single particle. However, this was not extensible to the many-particle case because it used an absolute time.

A renewed interest in constructing Lorentz-invariant extensions of Bohmian theory arose in the 1990s; see Bohm and Hiley: The Undivided Universe, and , , and references therein. Another approach is given in the work of Dürr et al., in which they use Bohm–Dirac models and a Lorentz-invariant foliation of space-time.

Thus, Dürr et al. (1999) showed that it is possible to formally restore Lorentz invariance for the Bohm–Dirac theory by introducing additional structure. This approach still requires a foliation of space-time. While this is in conflict with the standard interpretation of relativity, the preferred foliation, if unobservable, does not lead to any empirical conflicts with relativity. In 2013, Dürr et al. suggested that the required foliation could be covariantly determined by the wavefunction.

The relation between nonlocality and preferred foliation can be better understood as follows. In de Broglie–Bohm theory, nonlocality manifests as the fact that the velocity and acceleration of one particle depends on the instantaneous positions of all other particles. On the other hand, in the theory of relativity the concept of instantaneousness does not have an invariant meaning. Thus, to define particle trajectories, one needs an additional rule that defines which space-time points should be considered instantaneous. The simplest way to achieve this is to introduce a preferred foliation of space-time by hand, such that each hypersurface of the foliation defines a hypersurface of equal time.

Initially, it had been considered impossible to set out a description of photon trajectories in the de Broglie–Bohm theory in view of the difficulties of describing bosons relativistically. In 1996, Partha Ghose had presented a relativistic quantum-mechanical description of spin-0 and spin-1 bosons starting from the Duffin–Kemmer–Petiau equation, setting out Bohmian trajectories for massive bosons and for massless bosons (and therefore photons). In 2001, Jean-Pierre Vigier emphasized the importance of deriving a well-defined description of light in terms of particle trajectories in the framework of either the Bohmian mechanics or the Nelson stochastic mechanics. The same year, Ghose worked out Bohmian photon trajectories for specific cases. Subsequent weak-measurement experiments yielded trajectories that coincide with the predicted trajectories.

Chris Dewdney and G. Horton have proposed a relativistically covariant, wave-functional formulation of Bohm›s quantum field theory and have extended it to a form that allows the inclusion of gravity.

Nikolić has proposed a Lorentz-covariant formulation of the Bohmian interpretation of many-particle wavefunctions. He has developed a generalized relativistic-invariant probabilistic interpretation of quantum theory, in which $|\psi|^2$ is no longer a probability density in space, but a probability density in space-time. He uses this generalized probabilistic interpretation to formulate a relativistic-covariant version of de Broglie–Bohm theory without introducing a preferred foliation of space-time. His work also covers the extension of the Bohmian interpretation to a quantization of fields and strings.

Roderick I. Sutherland at the University in Sydney has a Lagrangian formalism for the pilot wave and its beables. I draws on Yakir Aharonov's retrocasual weak measurements to explain many-particle entanglement in a special relativistic way without the need for configuration space. The basic idea was already published by Costa de Beauregard in the 1950s and is also used by John Cramer in his transactional interpretation except the beables that exist between the von Neumann strong projection operator measurements. Sutherland's Lagrangian includes two-way action-reaction between pilot wave and beables. Therefore, it is a post-quantum non-statistical theory with final boundary conditions that violate the no-signal theorems of quantum theory. Just as special relativity is a limiting case of general relativity when the spacetime curvature vanishes, so, too is statistical no-entanglement signaling quantum theory with the Born rule a limiting case of the post-quantum action-reaction Lagrangian when the reaction is set to zero and the final boundary condition is integrated out.

Quantum Field Theory

In Dürr et al., the authors describe an extension of de Broglie–Bohm theory for handling creation and annihilation operators, which they refer to as "Bell-type quantum field theories". The basic idea is that configuration space becomes the (disjoint) space of all possible configurations of any number of particles. For part of the time, the system evolves deterministically under the guiding equation with a fixed number of particles. But under a stochastic process, particles may be created and annihilated. The distribution of creation events is dictated by the wavefunction. The wavefunction itself is evolving at all times over the full multi-particle configuration space.

Hrvoje Nikolić introduces a purely deterministic de Broglie–Bohm theory of particle creation and destruction, according to which particle trajectories are continuous, but particle detectors behave as if particles have been created or destroyed even when a true creation or destruction of particles does not take place.

Curved Space

To extend de Broglie–Bohm theory to curved space (Riemannian manifolds in mathematical parlance), one simply notes that all of the elements of these equations make sense, such as gradients and Laplacians. Thus, we use equations that have the same form as above. Topological and boundary conditions may apply in supplementing the evolution of Schrödinger's equation.

For a de Broglie–Bohm theory on curved space with spin, the spin space becomes a vector bundle over configuration space, and the potential in Schrödinger's equation becomes a local self-adjoint operator acting on that space.

Exploiting Nonlocality

The causal interpretation of quantum mechanics set up by de Broglie and Bohm was extended later by Bohm, Vigier, Hiley, Valentini and others to include stochastic properties. Bohm and other physicists, including Valentini, view the Born rule linking R to the probability density function $\rho = R^2$ as representing not a basic law, but rather as constituting a result of a system having reached *quantum equilibrium* during the course of the time development under the Schrödinger equation. It can be shown that, once an equilibrium has been reached, the system remains in such equilibrium over the course of its further evolution: this follows from the continuity equation associated with the Schrödinger evolution of ψ. However, it is less straightforward to demonstrate whether and how such an equilibrium is reached in the first place.

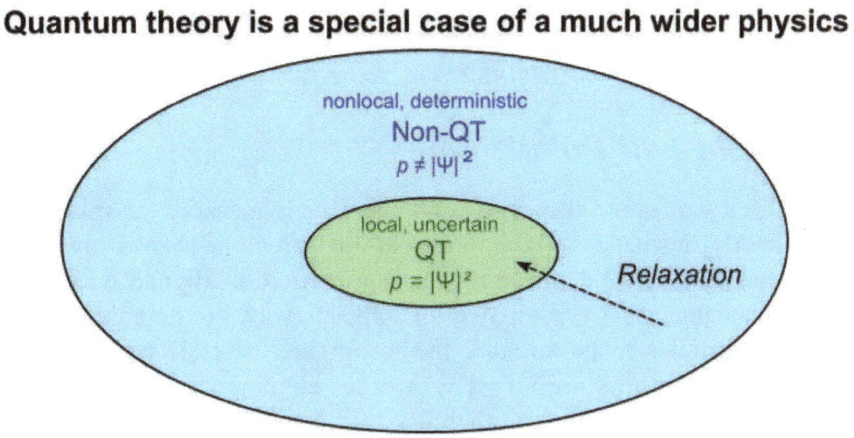

Diagram made by Antony Valentini in a lecture about the De Broglie–Bohm theory. Valentini argues quantum theory is a special equilibrium case of a wider physics and that it may be possible to observe and exploit quantum non-equilibrium

Antony Valentini has extended the de Broglie–Bohm theory to include signal nonlocality that would allow entanglement to be used as a stand-alone communication channel without a secondary classical "key" signal to "unlock" the message encoded in the entanglement. This violates orthodox quantum theory but has the virtue that it makes the parallel universes of the chaotic inflation theory observable in principle.

Unlike de Broglie–Bohm theory, Valentini›s theory has the wavefunction evolution also depending on the ontological variables. This introduces an instability, a feedback loop that pushes the hidden variables out of sub-quantal heat death. The resulting theory becomes nonlinear and non-unitary. Valentin argues that the laws of quantum mechanics are emergent and form a quantum equilibrium that has an analogous status to that of thermal equilibrium in classical dynamics. In principle therefore, other quantum non-equilibrium distributions may be potentially observed and exploited, for which the statistical predictions of quantum theory are violated. It is controversially argued that quantum theory is merely a special case of a much wider nonlinear physics, a physics in which non-local (superluminal) signalling is possible, and in which the uncertainty principle can be violated.

Results

Below are some highlights of the results that arise out of an analysis of de Broglie–Bohm theory. Experimental results agree with all of the standard predictions of quantum mechanics in so far as the latter has predictions. However, while standard quantum mechanics is limited to discussing the results of "measurements", de Broglie–Bohm theory is a theory that governs the dynamics of a system without the intervention of outside observers (p. 117 in Bell).

The basis for agreement with standard quantum mechanics is that the particles are distributed according to $|\psi|^2$. This is a statement of observer ignorance, but it can be proven that for a universe governed by this theory, this will typically be the case. There is apparent collapse of the wave function governing subsystems of the universe, but there is no collapse of the universal wavefunction.

Measuring Spin and Polarization

According to ordinary quantum theory, it is not possible to measure the spin or polarization of a particle directly; instead, the component in one direction is measured; the outcome from a single particle may be 1, meaning that the particle is aligned with the measuring apparatus, or −1, meaning that it is aligned the opposite way. For an ensemble of particles, if we expect the particles to be aligned, the results are all 1. If we expect them to be aligned oppositely, the results are all −1. For other alignments, we expect some results to be 1 and some to be −1 with a probability that depends on the expected alignment. For a full explanation of this, see the Stern–Gerlach experiment.

In de Broglie–Bohm theory, the results of a spin experiment cannot be analyzed without some knowledge of the experimental setup. It is possible to modify the setup so that the trajectory of the particle is unaffected, but that the particle with one setup registers as spin-up, while in the other setup it registers as spin-down. Thus, for the de Broglie–Bohm theory, the particle's spin is not an intrinsic property of the particle; instead spin is, so to speak, in the wavefunction of the particle in relation to the particular device being used to measure the spin. This is an illustration of what is sometimes referred to as contextuality and is related to naive realism about operators.

Measurements, the Quantum Formalism, and Observer Independence

De Broglie–Bohm theory gives the same results as quantum mechanics. It treats the wavefunction as a fundamental object in the theory, as the wavefunction describes how the particles move. This means that no experiment can distinguish between the two theories. This section outlines the ideas as to how the standard quantum formalism arises out of quantum mechanics. References include Bohm's original 1952 paper and Dürr et al.

Collapse of the Wavefunction

De Broglie–Bohm theory is a theory that applies primarily to the whole universe. That is, there is a single wavefunction governing the motion of all of the particles in the universe according to the guiding equation. Theoretically, the motion of one particle depends on the positions of all of the other particles in the universe. In some situations, such as in experimental systems, we can represent the system itself in terms of a de Broglie–Bohm theory in which the wavefunction of the system is obtained by conditioning on the environment of the system. Thus, the system can be analyzed with Schrödinger's equation and the guiding equation, with an initial $|\psi|^2$ distribution for the particles in the system (see the section on the conditional wavefunction of a subsystem for details).

It requires a special setup for the conditional wavefunction of a system to obey a quantum evolution. When a system interacts with its environment, such as through a measurement, the conditional wavefunction of the system evolves in a different way. The evolution of the universal wavefunction can become such that the wavefunction of the system appears to be in a superposition of distinct states. But if the environment has recorded the results of the experiment, then using the actual Bohmian configuration of the environment to condition on, the conditional wavefunction collapses to just one alternative, the one corresponding with the measurement results.

Collapse of the universal wavefunction never occurs in de Broglie–Bohm theory. Its entire evolution is governed by Schrödinger›s equation, and the particles› evolutions are governed by the guiding equation. Collapse only occurs in a phenomenological way for systems that seem to follow their own Schrödinger›s equation. As this is an effective description of the system, it is a matter of choice as to what to define the experimental system to include, and this will affect when collapse occurs.

Operators as Observables

In the standard quantum formalism, measuring observables is generally thought of as measuring operators on the Hilbert space. For example, measuring position is considered to be a measurement of the position operator. This relationship between physical measurements and Hilbert space operators is, for standard quantum mechanics, an additional axiom of the theory. The de Broglie–Bohm theory, by contrast, requires no such measurement axioms (and measurement as such is not a dynamically distinct or special sub-category of physical processes in the theory). In particular, the usual operators-as-observables formalism is, for de Broglie–Bohm theory, a theorem. A major point of the analysis is that many of the measurements of the observables do not correspond to properties of the particles; they are (as in the case of spin discussed above) measurements of the wavefunction.

In the history of de Broglie–Bohm theory, the proponents have often had to deal with claims that this theory is impossible. Such arguments are generally based on inappropriate analysis of operators as observables. If one believes that spin measurements are indeed measuring the spin of a particle that existed prior to the measurement, then one does reach contradictions. De Broglie–Bohm theory deals with this by noting that spin is not a feature of the particle, but rather that of the wavefunction. As such, it only has a definite outcome once the experimental apparatus is chosen. Once that is taken into account, the impossibility theorems become irrelevant.

There have also been claims that experiments reject the Bohm trajectories in favor of the standard QM lines. But as shown in and , such experiments cited above only disprove a misinterpretation of the de Broglie–Bohm theory, not the theory itself.

There are also objections to this theory based on what it says about particular situations usually involving eigenstates of an operator. For example, the ground state of hydrogen is a real wavefunction. According to the guiding equation, this means that the electron is at rest when in this state. Nevertheless, it is distributed according to $|\psi|^2$ and no contradiction to experimental results is possible to detect.

Operators as observables leads many to believe that many operators are equivalent. De Broglie–Bohm theory, from this perspective, chooses the position observable as a favored observable rather than, say, the momentum observable. Again, the link to the position observable is a consequence of the dynamics. The motivation for de Broglie–Bohm theory is to describe a system of particles. This implies that the goal of the theory is to describe the positions of those particles at all times. Other observables do not have this compelling ontological status. Having definite positions explains having definite results such as flashes on a detector screen. Other observables would not lead to that conclusion, but there need not be any problem in defining a mathematical theory for other observables; see Hyman et al. for an exploration of the fact that a probability density and probability current can be defined for any set of commuting operators.

Hidden Variables

De Broglie–Bohm theory is often referred to as a "hidden-variable" theory. Bohm used this description in his original papers on the subject, writing: "From the point of view of the usual interpretation, these additional elements or parameters could be called 'hidden' variables." Bohm and Hiley later stated that they found Bohm's choice of the term "hidden variables" to be too restrictive. In particular, they argued that a particle is not actually hidden but rather "is what is most directly manifested in an observation its properties cannot be observed with arbitrary precision (within the limits set by uncertainty principle)". However, others nevertheless treat the term "hidden variable" as a suitable description.

Generalized particle trajectories can be extrapolated from numerous weak measurements on an ensemble of equally prepared systems, and such trajectories coincide with the de Broglie–Bohm trajectories. In particular, an experiment with two entangled photons, in which a set of Bohmian trajectories for one of the photons was determined using weak measurements and postselection, can be understood in terms of a nonlocal connection between that photon's trajectory and the other photon's polarization. However, not only the De Broglie–Bohm interpretation, but also many other interpretation of quantum mechanics that do not include such trajectories are consistent with such experimental evidence.

Heisenberg's Uncertainty Principle

The Heisenberg's uncertainty principle states that when two complementary measurements are made, there is a limit to the product of their accuracy. As an example, if one measures the position with an accuracy of Δx and the momentum with an accuracy o Δp, then $\Delta x \Delta p \gtrsim h$. If we make further measurements in order to get more information, we disturb the system and change the trajectory into a new one depending on the measurement setup; therefore, the measurement results are still subject to Heisenberg's uncertainty relation.

In de Broglie–Bohm theory, there is always a matter of fact about the position and momentum of a particle. Each particle has a well-defined trajectory, as well as a wavefunction. Observers have limited knowledge as to what this trajectory is (and thus of the position and momentum). It is the lack of knowledge of the particle's trajectory that accounts for the uncertainty relation. What one can know about a particle at any given time is described by the wavefunction. Since the uncertainty relation can be derived from the wavefunction in other interpretations of quantum mechanics, it can be likewise derived (in the epistemic sense mentioned above) on the de Broglie–Bohm theory.

To put the statement differently, the particles' positions are only known statistically. As in classical mechanics, successive observations of the particles' positions refine the experimenter's knowledge of the particles' initial conditions. Thus, with succeeding observations, the initial conditions become more and more restricted. This formalism is consistent with the normal use of the Schrödinger equation.

For the derivation of the uncertainty relation, see Heisenberg uncertainty principle, noting that it describes it from the viewpoint of the Copenhagen interpretation.

Quantum Entanglement, Einstein–Podolsky–Rosen Paradox, Bell's Theorem, and Nonlocality

De Broglie–Bohm theory highlighted the issue of nonlocality: it inspired John Stewart Bell to prove his now-famous theorem, which in turn led to the Bell test experiments.

In the Einstein–Podolsky–Rosen paradox, the authors describe a thought experiment that one could perform on a pair of particles that have interacted, the results of which they interpreted as indicating that quantum mechanics is an incomplete theory.

Decades later John Bell proved Bell's theorem (see p. 14 in Bell), in which he showed that, if they are to agree with the empirical predictions of quantum mechanics, all such "hidden-variable" completions of quantum mechanics must either be nonlocal (as the Bohm interpretation is) or give up the assumption that experiments produce unique results (see counterfactual definiteness and many-worlds interpretation). In particular, Bell proved that any local theory with unique results must make empirical predictions satisfying a statistical constraint called "Bell's inequality".

Alain Aspect performed a series of Bell test experiments that test Bell's inequality using an EPR-type setup. Aspect's results show experimentally that Bell's inequality is in fact violated, meaning that the relevant quantum-mechanical predictions are correct. In these Bell test experiments, entangled pairs of particles are created; the particles are separated, traveling to remote measuring apparatus. The orientation of the measuring apparatus can be changed while the particles are in flight, demonstrating the apparent nonlocality of the effect.

The de Broglie–Bohm theory makes the same (empirically correct) predictions for the Bell test experiments as ordinary quantum mechanics. It is able to do this because it is manifestly nonlocal. It is often criticized or rejected based on this; Bell's attitude was: "It is a merit of the de Broglie–Bohm version to bring this out so explicitly that it cannot be ignored."

The de Broglie–Bohm theory describes the physics in the Bell test experiments as follows: to understand the evolution of the particles, we need to set up a wave equation for both particles; the orientation of the apparatus affects the wavefunction. The particles in the experiment follow the guidance of the wavefunction. It is the wavefunction that carries the faster-than-light effect of changing the orientation of the apparatus. An analysis of exactly what kind of nonlocality is present and how it is compatible with relativity can be found in Maudlin. Note that in Bell's work, and in more detail in Maudlin's work, it is shown that the nonlocality does not allow signaling at speeds faster than light.

Classical Limit

Bohm's formulation of de Broglie–Bohm theory in terms of a classically looking version has the merits that the emergence of classical behavior seems to follow immediately for any situation in which the quantum potential is negligible, as noted by Bohm in 1952. Modern methods of decoherence are relevant to an analysis of this limit. See Allori et al. for steps towards a rigorous analysis.

Quantum Trajectory Method

Work by Robert E. Wyatt in the early 2000s attempted to use the Bohm "particles" as an adaptive mesh that follows the actual trajectory of a quantum state in time and space. In the "quantum trajectory" method, one samples the quantum wavefunction with a mesh of quadrature points. One then evolves the quadrature points in time according to the Bohm equations of motion. At each time step, one then re-synthesizes the wavefunction from the points, recomputes the quantum forces, and continues the calculation. (QuickTime movies of this for H + H_2 reactive scattering can be found on the Wyatt group web-site at UT Austin.) This approach has been adapted, extended, and used by a number of researchers in the chemical physics community as a way to compute semi-classical and quasi-classical molecular dynamics. A recent (2007) issue of the Journal of Physical Chemistry A was dedicated to Prof. Wyatt and his work on "computational Bohmian dynamics.

Eric R. Bittner›s group at the University of Houston has advanced a statistical variant of this approach that uses Bayesian sampling technique to sample the quantum density and compute the quantum potential on a structureless mesh of points. This technique was recently used to estimate quantum effects in the heat capacity of small clusters Ne_n for $n \approx 100$.

There remain difficulties using the Bohmian approach, mostly associated with the formation of singularities in the quantum potential due to nodes in the quantum wavefunction. In general, nodes forming due to interference effects lead to the case where $R^{-1}\nabla^2 R \to \infty$. This results in an infinite force on the sample particles forcing them to move away from the node and often crossing the path of other sample points (which violates single-valuedness). Various schemes have been developed to overcome this; however, no general solution has yet emerged.

These methods, as does Bohm's Hamilton–Jacobi formulation, do not apply to situations in which the full dynamics of spin need to be taken into account.

Occam's-Razor Criticism

Both Hugh Everett III and Bohm treated the wavefunction as a physically real field.

Everett's many-worlds interpretation is an attempt to demonstrate that the wavefunction alone is sufficient to account for all our observations. When we see the particle detectors flash or hear the click of a Geiger counter, then Everett's theory interprets this as our *wavefunction* responding to changes in the detector's *wavefunction*, which is responding in turn to the passage of another *wavefunction* (which we think of as a "particle", but is actually just another wave packet). No particle (in the Bohm sense of having a defined position and velocity) exists, according to that theory. For this reason Everett sometimes referred to his own many-worlds approach as the "pure wave theory". Talking of Bohm's 1952 approach, Everett says:

In the Everettian view, then, the Bohm particles are superfluous entities, similar to, and equally as unnecessary as, for example, the luminiferous ether, which was found to be unnecessary in special relativity. This argument of Everett is sometimes called the redundancy argument, since the superfluous particles are redundant in the sense of Occam›s razor.

Many authors have expressed critical views of the de Broglie–Bohm theory by comparing it to Everett›s many-worlds approach. Many (but not all) proponents of the de Broglie–Bohm theory (such as Bohm and Bell) interpret the universal wavefunction as physically real. According to some supporters of Everett›s theory, if the (never collapsing) wavefunction is taken to be physically real, then it is natural to interpret the theory as having the same many worlds as Everett›s theory. In the Everettian view the role of the Bohmian particle is to act as a pointer, tagging, or selecting, just one branch of the universal wavefunction (the assumption that this branch indicates which *wave packet* determines the observed result of a given experiment is called the "result assumption"); the other branches are designated "empty" and implicitly assumed by Bohm to be devoid of conscious observers. H. Dieter Zeh comments on these "empty" branches:

According to Brown & Wallace, the de Broglie–Bohm particles play no role in the solution of the measurement problem. These authors claimthat the result assumption (see above) is inconsistent with the view that there is no measurement problem in the predictable outcome (i.e. single-outcome) case. These authors also claim that a standard tacit assumption of the de Broglie–Bohm theory (that an observer becomes aware of configurations of particles of ordinary objects by means of correlations between such configurations and the configuration of the particles in the observer›s brain) is unreasonable. This conclusion has been challenged by Valentini, who argues that the entirety of such objections arises from a failure to interpret de Broglie–Bohm theory on its own terms.

According to Peter R. Holland, in a wider Hamiltonian framework, theories can be formulated in which particles *do* act back on the wave function.

Non-Equivalence

In 2016, Pisin Chen and Hagen Kleinert argued that the Copenhagen interpretation and the De Broglie–Bohm theory yield different results for the ratio of peak intensities in the double-slit experiment. They concluded that they are thus not mathematically equivalent.

Derivations

De Broglie–Bohm theory has been derived many times and in many ways. Below are six derivations, all of which are very different and lead to different ways of understanding and extending this theory.

- Schrödinger's equation can be derived by using Einstein's light quanta hypothesis: $E = \hbar\omega$ and de Broglie's hypothesis $\mathbf{p} = \hbar\mathbf{k}$. The guiding equation can be derived in a similar fashion. We assume a plane wave: $\psi(\mathbf{x},t) = Ae^{i(\mathbf{k}\cdot\mathbf{x}-\omega t)}$. Notice that $i\mathbf{k} = \nabla\psi/\psi$. p = mv for the particle's actual velocity, we have that $\mathbf{v} = \frac{\hbar}{m}\mathrm{Im}\left(\frac{\nabla\psi}{\psi}\right)$. Thus, we have the guiding equation.
- Notice that this derivation does not use Schrödinger's equation.
- Preserving the density under the time evolution is another method of derivation. This is the method that Bell cites. It is this method that generalizes to many possible alternative theories. The starting point is the continuity equation $-\frac{\partial\rho}{\partial t} = \nabla\cdot(\rho v^{\psi})$ for the density $\rho = |\psi|^2$ This equation describes a probability flow along a current. We take the velocity field associated with this current as the velocity field whose integral curves yield the motion of the particle.
- A method applicable for particles without spin is to do a polar decomposition of the wavefunction and transform Schrödinger's equation into two coupled equations: the continuity equation from above and the Hamilton–Jacobi equation. This is the method used by Bohm in 1952. The decomposition and equations are as follows:

Decomposition: $\psi(\mathbf{x},t) = R(\mathbf{x},t)e^{iS(\mathbf{x},t)/\hbar}$. Note that $R^2(\mathbf{x},t)$ corresponds to the probability density $\rho(\mathbf{x},t) = |\psi(\mathbf{x},t)|^2$.

Continuity equation: $-\frac{\partial\rho(\mathbf{x},t)}{\partial t} = \nabla\cdot\left(\rho(\mathbf{x},t)\frac{\nabla S(\mathbf{x},t)}{m}\right)$.

Hamilton–Jacobi equation:

$$\frac{\partial S(\mathbf{x},t)}{\partial t} = -\left[\frac{1}{2m}(\nabla S(\mathbf{x},t))^2 + V - \frac{\hbar^2}{2m}\frac{\nabla^2 R(\mathbf{x},t)}{R(\mathbf{x},t)}\right].$$

The Hamilton–Jacobi equation is the equation derived from a Newtonian system with potential $V - \frac{\hbar^2}{2m}\frac{\nabla^2 R}{R}$ and velocity field $\frac{\nabla S}{m}$. The potential V is the classical potential that appears in Schrödinger's equation, and the other term involving R is the quantum potential, terminology introduced by Bohm.

This leads to viewing the quantum theory as particles moving under the classical force modified by a quantum force. However, unlike standard Newtonian mechanics, the initial velocity field is already specified by $\frac{\nabla S}{m}$ which is a symptom of this being a first-order theory, not a second-order theory.

- A fourth derivation was given by Dürr et al. In their derivation, they derive the velocity field by demanding the appropriate transformation properties given by the various symmetries that Schrödinger's equation satisfies, once the wavefunction is suitably transformed. The guiding equation is what emerges from that analysis.

- A fifth derivation, given by Dürr et al. is appropriate for generalization to quantum field theory and the Dirac equation. The idea is that a velocity field can also be understood as a first-order differential operator acting on functions. Thus, if we know how it acts on functions, we know what it is. Then given the Hamiltonian operator H, the equation to satisfy for all functions f (with associated multiplication operator $\hat{f}$) is $(v(f))(q) = \mathrm{Re}\,\frac{\left(\psi, \frac{i}{\hbar}[H, \hat{f}]\psi\right)}{(\psi, \psi)}(q)$, where (v, w) is the local Hermitian inner product on the value space of the wavefunction.
- This formulation allows for stochastic theories such as the creation and annihilation of particles.

A further derivation has been given by Peter R. Holland, on which he bases the entire work presented in his quantum-physics textbook *The Quantum Theory of Motion*, a main reference book on the de Broglie–Bohm theory. It is based on three basic postulates and an additional fourth postulate that links the wavefunction to measurement probabilities:

1. A physical system consists in a spatiotemporally propagating wave and a point particle guided by it.
2. The wave is described mathematically by a solution ψ to Schrödinger's wave equation.
3. The particle motion is described by a solution to $\dot{\mathbf{x}}(t) = [\nabla S(\mathbf{x}(t), t))]/m$ in dependence on initial condition $\mathbf{x}(t = 0)$, with S the phase of ψ The fourth postulate is subsidiary yet consistent with the first three:
4. The probability $\rho(\mathbf{x}(t))$ to find the particle in the differential volume d^3x time t equals $|\psi(\mathbf{x}(t))|^2$.

Pilot-Wave Theory

Louis de Broglie presented his pilot wave theory at the 1927 Solvay Conference, after close collaboration with Schrödinger, who developed his wave equation for de Broglie's theory. At the end of the presentation, Wolfgang Pauli pointed out that it was not compatible with a semi-classical technique Fermi had previously adopted in the case of inelastic scattering. Contrary to a popular legend, de Broglie actually gave the correct rebuttal that the particular technique could not be generalized for Pauli's purpose, although the audience might have been lost in the technical details and de Broglie's mild manner left the impression that Pauli's objection was valid. He was eventually persuaded to abandon this theory nonetheless because he was "discouraged by criticisms which roused". De Broglie's theory already applies to multiple spin-less particles, but lacks an adequate theory of measurement as no one understood quantum decoherence at the time. An analysis of de Broglie's presentation is given in Bacciagaluppi et al. Also, in 1932 John von Neumann published a paper, that was widely (and erroneously, as shown by Jeffrey Bub) believed to prove that all hidden-variable theories are impossible. This sealed the fate of de Broglie›s theory for the next two decades.

In 1926, Erwin Madelung had developed a hydrodynamic version of Schrödinger›s equation, which is incorrectly considered as a basis for the density current derivation of the de Broglie–Bohm theory. The Madelung equations, being quantum Euler equations (fluid dynamics), differ philosophically from the de Broglie–Bohm mechanics and are the basis of the stochastic interpretation of quantum mechanics.

Peter R. Holland has pointed out that, earlier in 1927, Einstein had actually submitted a preprint with a similar proposal but, not convinced, had withdrawn it before publication. According to Holland, failure to appreciate key points of the de Broglie–Bohm theory has led to confusion, the key point being that the trajectories of a many-body quantum system are correlated not because the particles exert a direct force on one another (*à la* Coulomb) but because all are acted upon by an entity – mathematically described by the wavefunction or functions of it – that lies beyond them. This entity is the quantum potential.

After publishing a popular textbook on Quantum Mechanics that adhered entirely to the Copenhagen orthodoxy, Bohm was persuaded by Einstein to take a critical look at von Neumann›s theorem. The result was ‹A Suggested Interpretation of the Quantum Theory in Terms of Hidden Variables I and II›. It was an independent origination of the pilot wave theory, and extended it to incorporate a consistent theory of measurement, and to address a criticism of Pauli that de Broglie did not properly respond to; it is taken to be deterministic (though Bohm hinted in the original papers that there should be disturbances to this, in the way Brownian motion disturbs Newtonian mechanics). This stage is known as the *de Broglie–Bohm Theory* in Bell's work and is the basis for 'The Quantum Theory of Motion'.

This stage applies to multiple particles, and is deterministic.

The de Broglie–Bohm theory is an example of a hidden variables theory. Bohm originally hoped that hidden variables could provide a local, causal, objective description that would resolve or eliminate many of the paradoxes of quantum mechanics, such as Schrödinger›s cat, the measurement problem and the collapse of the wavefunction. However, Bell›s theorem complicates this hope, as it demonstrates that there can be no local hidden variable theory that is compatible with the predictions of quantum mechanics. The Bohmian interpretation is causal but not local.

Bohm›s paper was largely ignored or panned by other physicists. Albert Einstein, who had suggested that Bohm search for a realist alternative to the prevailing Copenhagen approach, did not consider Bohm›s interpretation to be a satisfactory answer to the quantum nonlocality question, calling it too cheap, while Werner Heisenberg considered it a superfluous ‹ideological superstructure› . Wolfgang Pauli, who had been unconvinced by de Broglie in 1927, conceded to Bohm as follows:

I just received your long letter of 20th November, and I also have studied more thoroughly the details of your paper. I do not see any longer the possibility of any logical contradiction as long as your results agree completely with those of the usual wave mechanics and as long as no means is given to measure the values of your hidden parameters both in the measuring apparatus and in the observe system. As far as the whole matter stands now, your 'extra wave-mechanical predictions' are still a check, which cannot be cashed.

He subsequently described Bohm›s theory as artificial metaphysics.

According to physicist Max Dresden, when Bohm›s theory was presented at the Institute for Advanced Study in Princeton, many of the objections were ad hominem, focusing on Bohm›s sympathy with communists as exemplified by his refusal to give testimony to the House Un-American Activities Committee.

In 1979, Chris Philippidis, Chris Dewdney and Basil Hiley were the first to perform numeric computations on the basis of the quantum potential to deduce ensembles of particle trajectories. Their work renewed the interests of physicists in the Bohm interpretation of quantum physics.

Eventually John Bell began to defend the theory. In Speakable and Unspeakable in Quantum Mechanics several of the papers refer to hidden variables theories (which include Bohm›s).

The trajectories of the Bohm model that would result for particular experimental arrangements were termed surreal by some. Still in 2016, mathematical physicist Sheldon Goldstein said about Bohm›s theory: "There was a time when you couldn't even talk about it because it was heretical. It probably still is the kiss of death for a physics career to be actually working on Bohm, but maybe that's changing."

Bohmian Mechanics

This term is used to describe the same theory, but with an emphasis on the notion of current flow, which is determined on the basis of the quantum equilibrium hypothesis that the probability follows the Born rule. The term "Bohmian mechanics" is also often used to include most of the further extensions past the spin-less version of Bohm. While de Broglie–Bohm theory has Lagrangians and Hamilton-Jacobi equations as a primary focus and backdrop, with the icon of the quantum potential, Bohmian mechanics considers the continuity equation as primary and has the guiding equation as its icon. They are mathematically equivalent in so far as the Hamilton-Jacobi formulation applies, i.e., spin-less particles. The papers of Dürr et al. popularized the term.

All of non-relativistic quantum mechanics can be fully accounted for in this theory.

Causal Interpretation and Ontological Interpretation

Bohm developed his original ideas, calling them the *Causal Interpretation*. Later he felt that *causal* sounded too much like *deterministic* and preferred to call his theory the *Ontological Interpretation*. The main reference is "The Undivided Universe".

This stage covers work by Bohm and in collaboration with Jean-Pierre Vigier and Basil Hiley. Bohm is clear that this theory is non-deterministic (the work with Hiley includes a stochastic theory). As such, this theory is not, strictly speaking, a formulation of the de Broglie–Bohm theory. However, it deserves mention here because the term Bohm Interpretation is ambiguous between this theory and the de Broglie–Bohm theory.

An in-depth analysis of possible interpretations of Bohm›s model of 1952 was given in 1996 by philosopher of science Arthur Fine.

Hydrodynamic Quantum Analogs

Couder and Fort (2006) discovered that a millimetric droplet sustained on the surface of a vibrating fluid bath may self-propel through a resonant interaction with its own wave field. Quantization arises from the dynamic constraint imposed on the droplet by its pilot-wave field, and multimodal statistics appear to be a feature of chaotic pilot-wave dynamics.

Pioneering experiments on hydrodynamical analogs of quantum mechanics beginning with the work of Couder and Fort (2006) have shown that macroscopic classical pilot-waves can exhibit characteristics previously thought to be restricted to the quantum realm. Hydrodynamic pilot-wave analogs have been able to duplicate the double slit experiment, tunneling, quantized orbits, and numerous other quantum phenomena which have led to a resurgence in interest in pilot wave theories. Coulder and Fort note in their 2006 paper that pilot-waves are nonlinear dissipative systems sustained by external forces. A dissipative system is characterized by the spontaneous appearance of symmetry breaking (anisotropy) and the formation of complex, sometimes chaotic or emergent, dynamics where interacting fields can exhibit long range correlations. Stochastic electrodynamics (SED) an extension of the de Broglie–Bohm interpretation of quantum mechanics, with the electromagnetic zero-point field (ZPF) playing a central role as the guiding pilot-wave. Modern approaches to SED consider wave and particle-like quantum effects as well-coordinated emergent systems that are the result of speculated sub-quantum interactions with the zero-point field

A comparison by Bush (2015) among the walking droplet system, de Broglie's double-solution pilot-wave theory and its extension to SED

	Hydrodynamic walkers	de Broglie	SED pilot wave
Driving	bath vibration	internal clock	vacuum fluctuations
Spectrum	monochromatic	monochromatic	broad
Trigger	bouncing	zitterbewegung	zitterbewegung
Trigger frequency	ω_F	$\omega_c = mc^2/\hbar$	$\omega_c = mc^2/\hbar$
Energetics	GPE $\leftrightarrow$ wave	$mc^2 \leftrightarrow \hbar\omega$	$mc^2 \leftrightarrow$ EM
Resonance	droplet-wave	harmony of phases	unspecified
Dispersion $\omega(k)$	$\omega_F^2 \approx \sigma k^3/\rho$	$\omega^2 \approx \omega_c^2 + c^2k^2$	$\omega = ck$
Carrier λ	λ_F	λ_{dB}	λ_c
Statistical λ	λ_F	λ_{dB}	λ_{dB}

Experiments

Researchers performed the ESSW experiment. They found that the photon trajectories aren't surrealistic after all but more precisely, that the paths may seem surrealistic, but only if one fails to take into account the nonlocality inherent in Bohm's theory.

REFERENCE

1. Griffiths, Robert B. (1984). "Consistent Histories and the Interpretation of Quantum Mechanics". J. Stat. Phys. 35: 219.
2. Griffiths, Robert B. "The Consistent Histories Approach to Quantum Mechanics". Stanford Encyclopedia of Philosophy. Stanford University. Retrieved 2016-10-22.
3. F. Dowker and A. Kent, "Properties of Consistent Histories", Phys. Rev. Lett. 75, 3038 - 3041 (1995)
4. Omnès, Roland (1999). Understanding Quantum Mechanics. Princeton University Press. pp. 179, 257. ISBN 978-0-691-00435-8. LCCN 98042442.
5. A. Kent and J. McElwaine Quantum Prediction Algorithms Phys. Rev. A, 55, 1703--1720 (1997)
6. R. B. Griffiths, Consistent Quantum Theory, Cambridge University Press, 2003
7. R. Omnès, Quantum Philosophy, Princeton University Press, 1999. See part III, especially Chapter IX
8. Hermann Wimmel (1992). Quantum physics & observed reality: a critical interpretation of quantum mechanics. World Scientific. p. 2. ISBN 978-981-02-1010-6. Retrieved 9 May 2011.
9. J. Mehra and H. Rechenberg, The historical development of quantum theory, Springer-Verlag, 2001, p. 271.

10. Howard, Don (2004). "Who invented the Copenhagen Interpretation? A study in mythology". Philosophy of Science. 71 (5): 669–682. doi:10.1086/425941. JSTOR 10.1086/425941.

11. Bohm, David (1952). "A Suggested Interpretation of the Quantum Theory in Terms of "Hidden" Variables. I & II". Physical Review. 85 (2): 166–193. Bibcode:1952PhRv...85..166B. doi:10.1103/PhysRev.85.166.

12. H. Kragh, Quantum generations: A History of Physics in the Twentieth Century, Princeton University Press, 1999, p. 210. ("the term 'Copenhagen interpretation' was not used in the 1930s but first entered the physicist's vocabulary in 1955 when Heisenberg used it in criticizing certain unorthodox interpretations of quantum mechanics.")

13. Werner Heisenberg, Physics and Philosophy, Harper, 1958

14. Cramer, John G. (July 1986). "The Transactional Interpretation of Quantum Mechanics". Reviews of Modern Physics. 58 (3): 649. Bibcode:1986RvMP...58..647C. doi:10.1103/revmodphys.58.647.

15. "... there is no reason to consider these matter waves as less real than particles." Heisenberg, W. (1959/1971). Criticism and counterproposals to the Copenhagen interpretation of quantum theory, Chapter 8, pp. 114–128, in Physics and Philosophy: the Revolution in Modern Science, third impression 1971, George Allen & Unwin, London, at p. 118.

16. Bohr, N. (1928). 'The quantum postulate and the recent development of atomic theory', Nature, 121: 580–590, doi :10.1038/121580a0, p. 586: "there can be no question of an immediate connexion with our ordinary conceptions".

17. Heisenberg, W. (1959/1971). 'Language and reality in modern physics', Chapter 10, pp. 145–160, in Physics and Philosophy: the Revolution in Modern Science, George Allen & Unwin, London, ISBN 0-04-530016 X, p. 153: "our common concepts cannot be applied to the structure of the atoms."

18. Jammer, M. (1982). 'Einstein and quantum physics', pp. 59–76 in Albert Einstein: Historical and Cultural Perspectives; the Centennial Symposium in Jerusalem, edited by G. Holton, Y. Elkana, Princeton University Press, Princeton NJ, ISBN 0-691-08299-5. On pp. 73–74, Jammer quotes a 1952 letter from Einstein to Besso: "The present quantum theory is unable to provide the description of a real state of physical facts, but only of an (incomplete) knowledge of such. Moreover, the very concept of a real factual state is debarred by the orthodox theoreticians. The situation arrived at corresponds almost exactly to that of the good old Bishop Berkeley."

19. Heisenberg, W. (1927). Über den anschaulichen Inhalt der quantentheoretischen Kinematik und Mechanik, Z. Phys. 43: 172–198. Translation as ‹The actual content of quantum theoretical kinematics and mechanics› here: Since the statistical nature of quantum theory is so closely linked to the uncertainty in all observations or perceptions, one could be tempted to conclude that behind the observed, statistical world a real world is hidden, in which the law of causality is applicable. We want to state explicitly that we believe such speculations to be

both fruitless and pointless. The only task of physics is to describe the relation between observations.

20. Jammer, M. (1982). ‹Einstein and quantum physics›, pp. 59–76 in Albert Einstein: Historical and Cultural Perspectives; the Centennial Symposium in Jerusalem, edited by G. Holton, Y. Elkana, Princeton University Press, Princeton NJ, ISBN 0-691-08299-5, p. 72.
21. Belousek, D.W. (1996). Einstein›s 1927 unpublished hidden-variable theory: its background, context and significance. Stud. Hist. Phil. Mod. Phys. 21 (4): 431–461.
22. Holland, P (2005). What›s wrong with Einstein›s 1927 hidden-variable interpretation of quantum mechanics?. Foundations of Physics. 35 (2): 177–196. doi:10.1007/s10701-004-1940-7.
23. Born, M. (1955). Statistical interpretation of quantum mechanics. Science. 122 (3172): 675–679. Bibcode:1955Sci...122..675B. doi:10.1126/science.122.3172.675. PMID 17798674.
24. ... the statistical interpretation, which I have first suggested and which has been formulated in the most general way by von Neumann, ... Born, M. (1953). The interpretation of quantum mechanics, Brit. J. Phil. Sci., 4(14): 95–106.
25. Bohr, N. (1928). ‹The quantum postulate and the recent development of atomic theory›, Nature, 121: 580–590, doi :10.1038/121580a0, p. 586: In this connexion Born succeeded in obtaining a statistical interpretation of the wave functions, allowing a calculation of the probability of the individual transition processes required by the quantum postulate..
26. Ballentine, L.E. (1970). The statistical interpretation of quantum mechanics. Rev. Mod. Phys. 42 (4): 358–381. Bibcode:1970RvMP...42..358B. doi:10.1103/revmodphys.42.358.
27. Born, M. (1949). Einstein›s statistical theories, in Albert Einstein: Philosopher Scientist, ed. P.A. Schilpp, Open Court, La Salle IL, volume 1, pp. 161–177.
28. Of course the introduction of the observer must not be misunderstood to imply that some kind of subjective features are to be brought into the description of nature. Heisenberg, W. (1959/1971). Criticism and counterproposals to the Copenhagen interpretation of quantum theory, Chapter 8, pp. 114–128, in Physics and Philosophy: the Revolution in Modern Science, third impression 1971, George Allen & Unwin, London, at p. 121.
29. the collapse or reduction of the wave function. This was introduced by Heisenberg in his uncertainty paper and later postulated by von Neumann as a dynamical process independent of the Schrodinger equation, Claus Kiefer (2002). On the interpretation of quantum theory – from Copenhagen to the present day. arXiv:quant-ph/0210152 .
30. W. Heisenberg Über den anschaulichen Inhalt der quantentheoretischen Kinematik und Mechanik, Zeitschrift für Physik, Volume 43, 172–198 (1927), as translated by John Wheeler and Wojciech Zurek, in Quantum Theory and Measurement (1983), p. 74. (The determination of the position selects a definite q from the

totality of possibilities and limits the options for all subsequent measurements. ... The results of later measurements can only be calculated when one again ascribes to the electron a smaller wavepacket of extension λ (wavelength of the light used in the observation). Thus, every position determination reduces the wavepacket back to its original extension λ.)

31. David Bohm, A Suggested Interpretation of the Quantum Theory in Terms of Hidden Variables, I, Physical Review, (1952), 85, pp 166–179
32. David Bohm, A Suggested Interpretation of the Quantum Theory in Terms of Hidden Variables, II, Physical Review, (1952), 85, pp 180–193
33. Hugh Everett, Relative State Formulation of Quantum Mechanics, Reviews of Modern Physics vol 29, (1957) pp 454–462, based on unitary time evolution without discontinuities.
34. H. Dieter Zeh, On the Interpretation of Measurement in Quantum Theory, Foundation of Physics, vol. 1, pp. 69–76, (1970).
35. Wojciech H. Zurek, Pointer Basis of Quantum Apparatus: Into what Mixture does the Wave Packet Collapse?, Physical Review D, 24, pp. 1516–1525 (1981)
36. Wojciech H. Zurek, Environment-Induced Superselection Rules, Physical Review D, 26, pp.1862–1880, (1982)
37. Camilleri, K (2006). Heisenberg and the wave–particle duality. Stud. Hist. Phil. Mod. Phys. 37: 298–315.
38. Camilleri, K. (2009). Heisenberg and the Interpretation of Quantum Mechanics: the Physicist as Philosopher, Cambridge University Press, Cambridge UK, ISBN 978-0-521-88484-6.
39. Duane, W. (1923). The transfer in quanta of radiation momentum to matter, Proc. Natl. Acad. Sci. 9(5): 158–164.
40. Jammer, M. (1974). The Philosophy of Quantum Mechanics: the Interpretations of QM in Historical Perspective, Wiley, ISBN 0-471-43958-4, pp. 453–455.
41. Gribbin, J. Q for Quantum
42. Max Tegmark (1998). The Interpretation of Quantum Mechanics: Many Worlds or Many Words?. Fortsch. Phys. 46 (6–8): 855–862. arXiv:quant-ph/9709032 . Bibcode:1998ForPh..46..855T. doi:10.1002/(SICI)1521-3978(199811)46:6/8<855::AID-PROP855>3.0.CO;2-Q.
43. M. Schlosshauer; J. Kofler; A. Zeilinger (2013). A Snapshot of Foundational Attitudes Toward Quantum Mechanics. Studies in History and Philosophy of Science Part B: Studies in History and Philosophy of Modern Physics. 44 (3): 222–230. arXiv:1301.1069 . doi:10.1016/j.shpsb.2013.04.004.
44. C. Sommer, Another Survey of Foundational Attitudes Towards Quantum Mechanics, arXiv:1303.2719. http://arxiv.org/pdf/1303.2719v1
45. T. Norsen, S. Nelson, Yet Another Snapshot of Foundational Attitudes Toward Quantum Mechanics, arXiv:1306.4646. http://arxiv.org/pdf/1306.4646v2.pdf
46. Erwin Schrödinger, in an article in the Proceedings of the American Philosophical Society, 124, 323–38.

47. Nairz, Olaf; Brezger, Björn; Arndt, Markus; Zeilinger, Anton (2001). Diffraction of Complex Molecules by Structures Made of Light. Physical Review Letters. 87 (16): 160401. arXiv:quant-ph/0110012 . Bibcode:2001PhRvL..87p0401N. doi:10.1103/PhysRevLett.87.160401. PMID 11690188.
48. Brezger, Björn; Hackermüller, Lucia; Uttenthaler, Stefan; Petschinka, Julia; Arndt, Markus; Zeilinger, Anton (2002). Matter-Wave Interferometer for Large Molecules. Physical Review Letters. 88 (10): 100404. arXiv:quant-ph/0202158 . Bibcode:2002PhRvL..88j0404B. doi:10.1103/PhysRevLett.88.100404. PMID 11909334.
49. Michael price on nonlocality in Many Worlds
50. Relativity and Causality in the Transactional Interpretation
51. Collapse and Nonlocality in the Transactional Interpretation
52. Werner Heisenberg, Physics and Philosophy, Harper, 1958, p. 137.
53. God does not throw dice quote
54. A. Pais, Einstein and the quantum theory, Reviews of Modern Physics 51, 863–914 (1979), p. 907.
55. Since the Universe naturally contains all of its observers, the problem arises to come up with an interpretation of quantum theory that contains no classical realms on the fundamental level.›, Claus Kiefer (2002). On the interpretation of quantum theory – from Copenhagen to the present day. arXiv:quant-ph/0210152 .
56. Jaynes, E. T. (1989). Clearing up Mysteries – The Original Goal(PDF). Maximum Entropy and Bayesian Methods: 7.
57. Momentum Transfer to a Free Floating Double Slit: Realization of a Thought Experiment from the Einstein-Bohr Debates, L. Ph. H. Schmidt et al. Physical Review Letters Week ending 2013
58. More correctly, when the law of large numbers is applied to solve this problem (so that the opposite change must also occur), a deterministic ensemble interpretation follows from the same law.
59. Schmidt, L.P.H., Lower, J., Jahnke, T., Schößler, S., Schöffler, M.S., Menssen, A., Lévêque, C., Sisourat, N., Taïeb, R., Schmidt-Böcking, H., Dörner, R. (2013). Momentum transfer to a free floating double slit: realization of a thought experiment from the Einstein–Bohr debates, Physical Review Letters 111: 103201, 1–5. See also the article on Bohr–Einstein debates. Likely there are even more such apparent interactions in various areas of the photon, for example when reflecting from the whole shutter.
60. Heisenberg, W. (1930). The Physical Principles of the Quantum Theory, translated by C. Eckart and F.C. Hoyt, University of Chicago Press, Chicago, pp. 77–78.
61. Pauling, L.C., Wilson, E.B. (1935). Introduction to Quantum Mechanics: with Applications to Chemistry, McGraw-Hill, New York, pp. 34–36.
62. Landé, A. (1951). Quantum Mechanics, Sir Isaac Pitman and Sons, London, pp. 19–22.
63. Van Vliet, K. (1967). Linear momentum quantization in periodic structures. Physica.

35: 97–106. Bibcode:1967Phy....35...97V. doi:10.1016/0031-8914(67)90138-3.

64. Van Vliet, K. (2010). Linear momentum quantization in periodic structures ii. Physica A. 389 (8): 1585–1593. Bibcode:2010PhyA..389.1585V. doi:10.1016/j.physa.2009.12.026.
65. Kate Becker (2013-01-25). Quantum physics has been rankling scientists for decades. Boulder Daily Camera. Retrieved 2013-01-25.
66. Schlosshauer, Maximilian; Kofler, Johannes; Zeilinger, Anton (2013-01-06). A Snapshot of Foundational Attitudes Toward Quantum Mechanics. Studies in History and Philosophy of Science Part B: Studies in History and Philosophy of Modern Physics. 44 (3): 222–230. arXiv:1301.1069 . doi:10.1016/j.shpsb.2013.04.004.
67. The Quantum Liar Experiment, RE Kastner, Studies in History and Philosophy of Modern Physics, Vol41, Iss.2,May2010
68. The non-relativistic Schrödinger equation does not admit advanced solutions.
69. http://home.fnal.gov/~skands/slides/A-Quantum-Journey.ppt
70. N. David Mermin (2004). Could Feynman Have Said This?. Physics Today. 57 (5): 10–11. Bibcode:2004PhT....57e..10M. doi:10.1063/1.1768652.

Quantum Gravity

Chapter Contents

Quantum Gravity is the name given to any theory that describes *gravity* in the regimes where *quantum* effects cannot be disregarded. At present, there is no such a theory which is universally accepted and confirmed by experience. Therefore the term "Quantum Gravity" indicates more an open problem than a specific theory. Several research lines, at different levels of development, offer tentative solutions to the problem.

INTRODUCTION

Quantum gravity (QG) is a field of theoretical physics that seeks to describe gravity according to the principles of quantum mechanics, and where quantum effects cannot be ignored.

The current understanding of gravity is based on Albert Einstein's general theory of relativity, which is formulated within the framework of classical physics. On the other hand, the nongravitational forces are described within the framework of quantum mechanics, a radically different formalism for describing physical phenomena based on the wave-like nature of matter. The necessity of a quantum mechanical description of gravity follows from the fact that one cannot consistently couple a classical system to a quantum one.

Although a quantum theory of gravity is needed in order to reconcile general relativity with the principles of quantum mechanics, difficulties arise when one attempts to apply the usual prescriptions of quantum field theory to the force of gravity via graviton bosons. The problem is that the theory one gets in this way is not renormalizable and therefore cannot be used to make meaningful physical predictions. As a result, theorists have taken up more radical approaches to the problem of quantum gravity, the most popular approaches being string theory and loop quantum gravity. A recent development is the theory of causal fermion systems which gives quantum mechanics, general relativity, and quantum field theory as limiting cases.

Strictly speaking, the aim of quantum gravity is only to describe the quantum behavior of the gravitational field and should not be confused with the objective of unifying all fundamental interactions into a single mathematical framework. While any substantial improvement into the present understanding of gravity would aid further work towards unification, study of quantum gravity is a field in its own right with various branches having different approaches to unification. Although some quantum gravity theories, such as string theory, try to unify gravity with the other fundamental forces, others, such as loop quantum gravity, make no such attempt; instead, they make an effort to quantize the gravitational field while it is kept separate from the other forces. A theory of quantum gravity that is also a grand unification of all known interactions is sometimes referred to as The Theory of Everything (TOE).

One of the difficulties of quantum gravity is that quantum gravitational effects are only expected to become apparent near the Planck scale, a scale far smaller in distance (equivalently, far larger in energy) than what is currently accessible at high energy particle accelerators. As a result, quantum gravity is a mainly theoretical enterprise, although there are speculations about how quantum gravitational effects might be observed in existing experiments.

OVERVIEW

Much of the difficulty in meshing these theories at all energy scales comes from the different assumptions that these theories make on how the universe works. Quantum field theory depends on particle fields embedded in the flat space-time of special relativity. General relativity models gravity as a curvature within space-time that changes as a gravitational mass moves. Historically, the most obvious way of combining the two (such as treating gravity as simply another particle field) ran quickly into what is known as the renormalization problem. In the old-fashioned understanding of renormalization, gravity particles would attract each other and adding together all of the interactions results in many infinite values which cannot easily be cancelled out mathematically to yield sensible, finite results. This is in contrast with quantum electrodynamics where, given that the series still do not converge, the interactions sometimes evaluate to infinite results, but those are few enough in number to be removable via renormalization.

Effective Field Theories

Quantum gravity can be treated as an effective field theory. Effective quantum field theories come with some high-energy cutoff, beyond which we do not expect that the theory provides a good description of nature. The "infinities" then become large but finite quantities depending on this finite cutoff scale, and correspond to processes that involve very high energies near the fundamental cutoff. These quantities can then be absorbed into an infinite collection of coupling constants, and at energies well below the fundamental cutoff of the theory, to any desired precision; only a finite number of these coupling constants need to be measured in order to make legitimate quantum-mechanical predictions. This same logic works just as well for the highly successful theory of low-energy pions as for quantum gravity. Indeed, the first quantum-mechanical corrections to graviton-scattering and Newton's law of gravitation have been explicitly computed (although they are so infinitesimally small that we may never be able to measure them). In fact, gravity is in many ways a much better quantum field theory than the Standard Model, since it appears to be valid all the way up to its cutoff at the Planck scale.

While confirming that quantum mechanics and gravity are indeed consistent at reasonable energies, it is clear that near or above the fundamental cutoff of our effective quantum theory of gravity (the cutoff is generally assumed to be of the order of the Planck scale), a new model of nature will be needed. Specifically, the problem of combining quantum mechanics and gravity becomes an issue only at very high energies, and may well require a totally new kind of model.

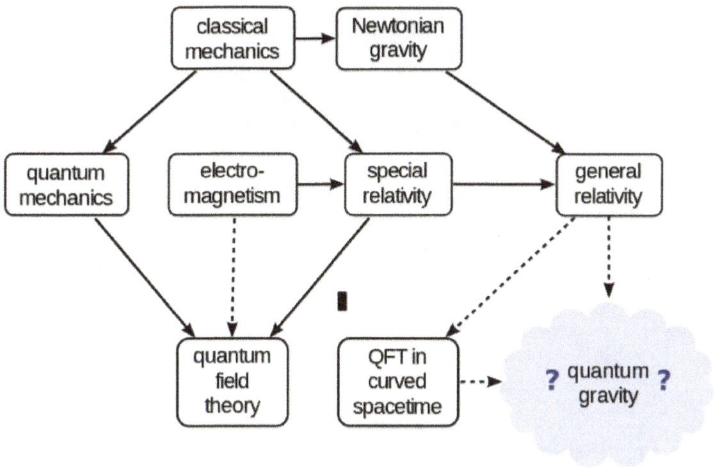

Diagram showing where quantum gravity sits in the hierarchy of physics theories

Quantum Gravity Theory for the Highest Energy Scales

The general approach to deriving a quantum gravity theory that is valid at even the highest energy scales is to assume that such a theory will be simple and elegant and, accordingly, to study symmetries and other clues offered by current theories that might suggest ways to combine them into a comprehensive, unified theory. One problem with this approach is that it is unknown whether quantum gravity will actually conform to a simple and elegant theory, as it should resolve the dual conundrums of special relativity with regard to the uniformity of acceleration and gravity, and general relativity with regard to spacetime curvature.

Such a theory is required in order to understand problems involving the combination of very high energy and very small dimensions of space, such as the behavior of black holes, and the origin of the universe.

QUANTUM MECHANICS AND GENERAL RELATIVITY

Gravity Probe B (GP-B) has measured spacetime curvature near Earth to test related models in application of Einstein's general theory of relativity.

Graviton

At present, one of the deepest problems in theoretical physics is harmonizing the theory of general relativity, which describes gravitation, and applications to large-scale structures (stars, planets, galaxies), with quantum mechanics, which describes the other three fundamental forces acting on the atomic scale. This problem must be put in the proper context, however. In particular, contrary to the popular claim that quantum mechanics and general relativity are fundamentally incompatible, one can demonstrate that the structure of general relativity essentially follows inevitably from the quantum mechanics of interacting theoretical spin-2 massless particles (called gravitons).

While there is no concrete proof of the existence of gravitons, quantized theories of matter may necessitate their existence. Supporting this theory is the observation that all fundamental forces except gravity have one or more known messenger particles, leading researchers to believe that at least one most likely does exist; they have dubbed this hypothetical particle the *graviton*. The predicted find would result in the classification of the graviton as a "force particle" similar to the photon of the electromagnetic field. Many of the accepted notions of a unified theory of physics since the 1970s assume, and to some degree depend upon, the existence of the graviton. These include string theory, superstring theory, M-theory, and loop quantum gravity. Detection of gravitons is thus vital to the validation of various lines of research to unify quantum mechanics and relativity theory.

Dilaton

The dilaton made its first appearance in Kaluza–Klein theory, a five-dimensional theory that combined gravitation and electromagnetism. Generally, it appears in string theory. More recently, however, it›s become central to the lower-dimensional many-bodied gravity problem based on the field theoretic approach of Roman Jackiw. The impetus arose from the fact that complete analytical solutions for the metric of a covariant *N*-body system have proven elusive in general relativity. To simplify the problem, the number of dimensions was lowered to *(1+1)*, i.e., one spatial dimension and one temporal dimension. This model problem, known as *R*=*T* theory (as opposed to the general *G*=*T* theory) was amenable to exact solutions in terms of a generalization of the Lambert W function. It was also found that the field equation governing the dilaton (derived from differential geometry) was the Schrödinger equation and consequently amenable to quantization.

Thus, one had a theory which combined gravity, quantization, and even the electromagnetic interaction, promising ingredients of a fundamental physical theory. It is worth noting that this outcome revealed a previously unknown and already existing *natural link* between general relativity and quantum mechanics. For some time, a generalization of this theory to *(3+1)* dimensions was unclear. However, a recent derivation in *(3+1)* dimensions under the right coordinate conditions yields a formulation similar to the earlier *(1+1)* namely a dilaton field governed by the logarithmic Schrödinger equation which is seen in condensed matter physics and superfluids. The field equations are indeed amenable to such a generalization (as shown with the inclusion of a one-graviton process) and yield the correct Newtonian limit in *d* dimensions but only if a dilaton is included. Furthermore, the results become even more tantalizing in view of the apparent resemblance between the dilaton and the Higgs boson. However, more experimentation is needed to resolve the relationship between these two particles. Since this theory can combine gravitational, electromagnetic and quantum effects, their coupling could potentially lead to a means of vindicating the theory, through cosmology and even, perhaps, experimentally.

Nonrenormalizability of Gravity

General relativity, like electromagnetism, is a classical field theory. One might expect that, as with electromagnetism, the gravitational force should also have a corresponding quantum field theory.

However, gravity is perturbatively nonrenormalizable. For a quantum field theory to be well-defined according to this understanding of the subject, it must be asymptotically free or asymptotically safe. The theory must be characterized by a choice of *finitely many* parameters, which could, in principle, be set by experiment. For example, in quantum electrodynamics these parameters are the charge and mass of the electron, as measured at a particular energy scale.

On the other hand, in quantizing gravity there are, in perturbation theory, *infinitely many independent parameters* (counterterm coefficients) needed to define the theory. For a given choice of those parameters, one could make sense of the theory, but since it›s impossible to conduct infinite experiments to fix the values of every parameter, it has been argued that one does not, in perturbation theory, have a meaningful physical theory:

- At low energies, the logic of the renormalization group tells us that, despite the unknown choices of these infinitely many parameters, quantum gravity will reduce to the usual Einstein theory of general relativity.
- On the other hand, if we could probe very high energies where quantum effects take over, then *every one* of the infinitely many unknown parameters would begin to matter, and we could make no predictions at all.

If we treat QG as an effective field theory, there is a way around this problem.

That is, the meaningful theory of quantum gravity (that makes sense and is predictive at all energy levels) inherently implies some deep principle that reduces the infinitely many unknown parameters to a finite number that can then be measured:

- One possibility is that normal perturbation theory is not a reliable guide to the renormalizability of the theory, and that there really *is* a UV fixed point for gravity. Since this is a question of non-perturbative quantum field theory, it is difficult to find a reliable answer, but some people still pursue this option.
- Another possibility is that there are new unfound symmetry principles that constrain the parameters and reduce them to a finite set. This is the route taken by string theory, where all of the excitations of the string essentially manifest themselves as new symmetries.

QG as an Effective Field Theory

In an effective field theory, all but the first few of the infinite set of parameters in a non-renormalizable theory are suppressed by huge energy scales and hence can be neglected when computing low-energy effects. Thus, at least in the low-energy regime, the model is indeed a predictive quantum field theory. (A very similar situation occurs for the very similar effective field theory of low-energy pions.) Furthermore, many theorists agree that even the Standard Model should really be regarded as an effective field theory as well, with "nonrenormalizable" interactions suppressed by large energy scales and whose effects have consequently not been observed experimentally.

Recent work has shown that by treating general relativity as an effective field theory, one can actually make legitimate predictions for quantum gravity, at least for low-energy phenomena. An example is the well-known calculation of the tiny first-order quantum-mechanical correction to the classical Newtonian gravitational potential between two masses.

Spacetime Background Dependence

A fundamental lesson of general relativity is that there is no fixed spacetime background, as found in Newtonian mechanics and special relativity; the spacetime geometry is dynamic. While easy to grasp in principle, this is the hardest idea to understand about general relativity, and its consequences are profound and not fully explored, even at the classical level. To a certain extent, general relativity can be seen to be a relational theory, in which the only physically relevant information is the relationship between different events in space-time.

On the other hand, quantum mechanics has depended since its inception on a fixed background (non-dynamic) structure. In the case of quantum mechanics, it is time that is given and not dynamic, just as in Newtonian classical mechanics. In relativistic quantum field theory, just as in classical field theory, Minkowski spacetime is the fixed background of the theory.

String Theory

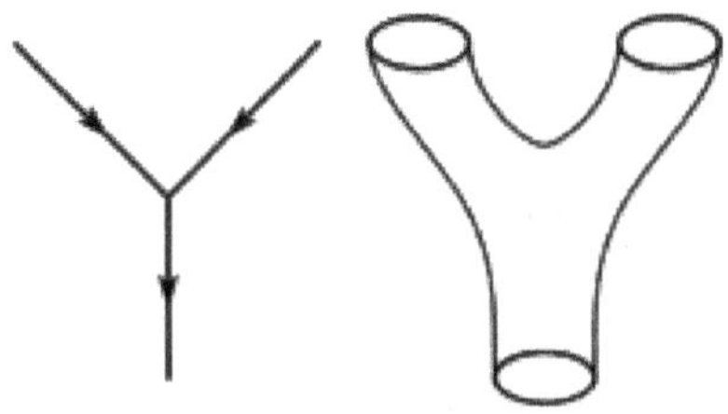

Interaction in the subatomic world: world lines of point-like particles in the Standard Model or a world sheet swept up by closed strings in string theory

String theory can be seen as a generalization of quantum field theory where instead of point particles, string-like objects propagate in a fixed spacetime background, although the interactions among closed strings give rise to space-time in a dynamical way. Although string theory had its origins in the study of quark confinement and not of quantum gravity, it was soon discovered that the string spectrum contains the graviton, and that "condensation" of certain vibration modes of strings is equivalent to a modification of the original background. In this sense, string perturbation theory exhibits exactly the features one would expect of a perturbation theory that may exhibit a strong dependence on asymptotics which is a weak form of background dependence.

Background Independent Theories

Loop quantum gravity is the fruit of an effort to formulate a background-independent quantum theory.

Topological quantum field theory provided an example of background-independent quantum theory, but with no local degrees of freedom, and only finitely many degrees of freedom globally. This is inadequate to describe gravity in 3+1 dimensions, which has local degrees of freedom according to general relativity. In 2+1 dimensions, however, gravity is a topological field theory, and it has been successfully quantized in several different ways, including spin networks.

Semi-Classical Quantum Gravity

Quantum field theory on curved (non-Minkowskian) backgrounds, while not a full quantum theory of gravity, has shown many promising early results. In an analogous way to the development of quantum electrodynamics in the early part of the 20th century (when physicists considered quantum mechanics in classical electromagnetic fields), the consideration of quantum field theory on a curved background has led to predictions such as black hole radiation.

Phenomena such as the Unruh effect, in which particles exist in certain accelerating frames but not in stationary ones, do not pose any difficulty when considered on a curved background (the Unruh effect occurs even in flat Minkowskian backgrounds). The vacuum state is the state with the least energy (and may or may not contain particles).

Problem of Time

A conceptual difficulty in combining quantum mechanics with general relativity arises from the contrasting role of time within these two frameworks. In quantum theories time acts as an independent background through which states evolve, with the Hamiltonian operator acting as the generator of infinitesmal translations of quantum states through time. In contrast, general relativity treats time as a dynamical variable which interacts directly with matter and moreover requires the Hamiltonian constraint to vanish, removing any possibility of employing a notion of time similar to that in quantum theory.

CANDIDATE THEORIES

There are a number of proposed quantum gravity theories. Currently, there is still no complete and consistent quantum theory of gravity, and the candidate models still need to overcome major formal and conceptual problems. They also face the common problem that, as yet, there is no way to put quantum gravity predictions to experimental tests, although there is hope for this to change as future data from cosmological observations and particle physics experiments becomes available.

String Theory

One suggested starting point is ordinary quantum field theories which, after all, are successful in describing the other three basic fundamental forces in the context of the standard model of elementary particle physics. However, while this leads to an acceptable effective (quantum) field theory of gravity at low energies, gravity turns out to be much more problematic at higher energies. For ordinary field theories such as quantum electrodynamics, a technique known as renormalization is an integral part of deriving predictions which take into account higher-energy contributions, but gravity turns out to be nonrenormalizable: at high energies, applying the recipes of ordinary quantum field theory yields models that are devoid of all predictive power.

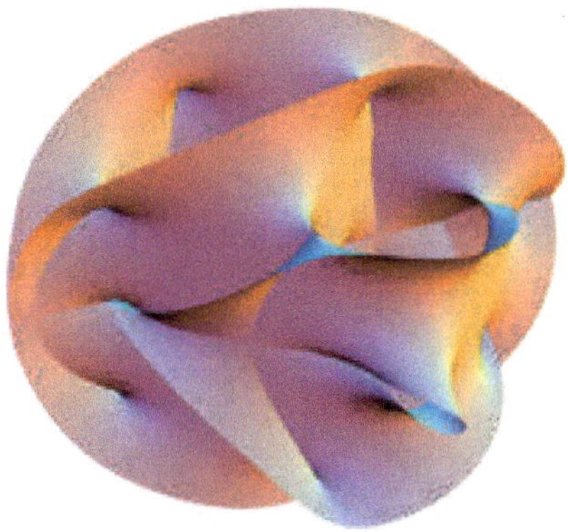

Projection of a Calabi–Yau manifold, one of the ways of compactifying the extra dimensions posited by string theory

One attempt to overcome these limitations is to replace ordinary quantum field theory, which is based on the classical concept of a point particle, with a quantum theory of one-dimensional extended objects: string theory. At the energies reached in current experiments, these strings are indistinguishable from point-like particles, but, crucially, different modes of oscillation of one and the same type of fundamental string appear as particles with different (electric and other) charges. In this way, string theory promises to be a unified description of all particles and interactions. The theory is successful in that one mode will always correspond to a graviton, the messenger particle of gravity; however, the price of this success are unusual features such as six extra dimensions of space in addition to the usual three for space and one for time.

In what is called the second superstring revolution, it was conjectured that both string theory and a unification of general relativity and supersymmetry known as supergravity form part of a hypothesized eleven-dimensional model known as M-theory, which would constitute a uniquely defined and consistent theory of quantum gravity. As presently understood, however, string theory admits a very large number (10^{500} by some estimates) of consistent vacua, comprising the so-called «string landscape». Sorting through this large family of solutions remains a major challenge.

Loop Quantum Gravity

Loop quantum gravity seriously considers general relativity's insight that spacetime is a dynamical field and is therefore a quantum object. Its second idea is that the quantum discreteness that determines the particle-like behavior of other field theories (for instance, the photons of the electromagnetic field) also affects the structure of space.

The main result of loop quantum gravity is the derivation of a granular structure of space at the Planck length. This is derived from following considerations: In the case of electromagnetism, the quantum operator representing the energy of each frequency of the field has a discrete spectrum. Thus the energy of each frequency is quantized, and the quanta are the photons. In the case of gravity, the operators representing the area and the volume of each surface or space region likewise have discrete spectrum. Thus area and volume of any portion of space are also quantized, where the quanta are elementary quanta of space. It follows, then, that spacetime has an elementary quantum granular structure at the Planck scale, which cuts off the ultraviolet infinities of quantum field theory.

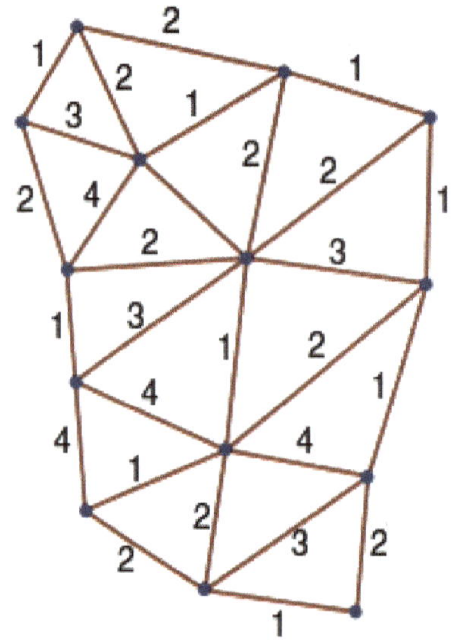

Simple spin network of the type used in loop quantum gravity

The quantum state of spacetime is described in the theory by means of a mathematical structure called spin networks. Spin networks were initially introduced by Roger Penrose in abstract form, and later shown by Carlo Rovelli and Lee Smolin to derive naturally from a non-perturbative quantization of general relativity. Spin networks do not represent quantum states of a field in spacetime: they represent directly quantum states of spacetime.

The theory is based on the reformulation of general relativity known as Ashtekar variables, which represent geometric gravity using mathematical analogues of electric and magnetic fields. In the quantum theory, space is represented by a network structure called a spin network, evolving over time in discrete steps.

The dynamics of the theory is today constructed in several versions. One version starts with the canonical quantization of general relativity. The analogue of the Schrödinger equation is a Wheeler–DeWitt equation, which can be defined within the theory. In the covariant, or spinfoam formulation of the theory, the quantum dynamics is obtained via a sum over discrete versions of spacetime, called spinfoams. These represent histories of spin networks.

Other Approaches

There are a number of other approaches to quantum gravity. The approaches differ depending on which features of general relativity and quantum theory are accepted unchanged, and which features are modified. Examples include:

- Asymptotic safety in quantum gravity
- Euclidean quantum gravity
- Causal dynamical triangulation
- Causal fermion systems, giving quantum mechanics, general relativity and quantum field theory as limiting cases.
- Causal sets
- Covariant Feynman path integral approach
- Group field theory
- Wheeler–DeWitt equation
- Geometrodynamics
- Hořava–Lifshitz gravity
- MacDowell–Mansouri action
- Path-integral based models of quantum cosmology
- Regge calculus
- Scale relativity
- Shape Dynamics
- String-nets giving rise to gapless helicity ±2 excitations with no other gapless excitations
- Superfluid vacuum theory a.k.a. theory of BEC vacuum
- Supergravity
- Twistor theory
- Canonical quantum gravity
- E8 Theory
- Quantum holonomy theory

Weinberg–Witten Theorem

In quantum field theory, the Weinberg–Witten theorem places some constraints on theories of composite gravity/emergent gravity. However, recent developments attempt to show that if locality is only approximate and the holographic principle is correct, the Weinberg–Witten theorem would not be valid.

EXPERIMENTAL TESTS

As was emphasized above, quantum gravitational effects are extremely weak and therefore difficult to test. For this reason, the possibility of experimentally testing quantum gravity had not received much attention prior to the late 1990s. However, in the past decade, physicists have realized that evidence for quantum gravitational effects can guide the development of the theory. Since theoretical development has been slow, the field of phenomenological quantum gravity, which studies the possibility of experimental tests, has obtained increased attention.

The most widely pursued possibilities for quantum gravity phenomenology include violations of Lorentz invariance, imprints of quantum gravitational effects in the cosmic microwave background (in particular its polarization), and decoherence induced by fluctuations in the space-time foam.

The BICEP2 experiment detected what was initially thought to be primordial B-mode polarization caused by gravitational waves in the early universe. If truly primordial, these waves were born as quantum fluctuations in gravity itself. Cosmologist Ken Olum (Tufts University) stated: "I think this is the only observational evidence that we have that actually shows that gravity is quantized....It's probably the only evidence of this that we will ever have."

REFERENCES

1. Rovelli, Carlo. "Quantum gravity - Scholarpedia". www.scholarpedia.org. Retrieved 2016-01-09.
2. Griffiths, David J. (2004). Introduction to Quantum Mechanics. Pearson Prentice Hall. OCLC 803860989.
3. Wald, Robert M. (1984). General Relativity. University of Chicago Press. p. 382. OCLC 471881415.
4. Zee, Anthony (2010). Quantum Field Theory in a Nutshell (2nd ed.). Princeton University Press. p. 172. OCLC 659549695.
5. Penrose, Roger (2007). The road to reality : a complete guide to the laws of the universe. Vintage. p. 1017. OCLC 716437154.
6. F. Finster, J. Kleiner, Causal fermion systems as a candidate for a unified physical theory, arXiv:1502.03587 math-ph (2015)
7. F. Finster, The Principle of the Fermionic Projector, hep-th/0001048, hep-th/0202059, hep- th/0210121, AMS/IP Studies in Advanced Mathematics, vol. 35, American Mathematical Society, Providence, RI, 2006.
8. F. Finster, A formulation of quantum field theory realizing a sea of interacting Dirac particles, arXiv:0911.2102 hep-th, Lett. Math. Phys. 97 (2011), no. 2, 165–183.
9. F. Finster, An action principle for an interacting fermion system and its analysis in the continuum limit, arXiv:0908.1542 math-ph (2009).
10. F. Finster, The continuum limit of a fermion system involving neutrinos: Weak

and gravitational interactions, arXiv:1211.3351 math-ph (2012).

11. F. Finster, Perturbative quantum field theory in the framework of the fermionic projector, arXiv:1310.4121 math-ph, J. Math. Phys. 55 (2014), no. 4, 042301.
12. Quantum effects in the early universe might have an observable effect on the structure of the present universe, for example, or gravity might play a role in the unification of the other forces. Cf. the text by Wald cited above.
13. Donoghue (1995). "Introduction to the Effective Field Theory Description of Gravity". arXiv:gr-qc/9512024 . (verify against ISBN 9789810229085)
14. Kraichnan, R. H. (1955). "Special-Relativistic Derivation of Generally Covariant Gravitation Theory". Physical Review. 98 (4): 1118–1122. Bibcode:1955PhRv...98.1118K. doi:10.1103/PhysRev.98.1118.
15. Gupta, S. N. (1954). "Gravitation and Electromagnetism". Physical Review. 96 (6): 1683–1685. Bibcode:1954PhRv...96.1683G. doi:10.1103/PhysRev.96.1683.
16. Gupta, S. N. (1957). "Einstein's and Other Theories of Gravitation". Reviews of Modern Physics. 29 (3): 334–336. Bibcode:1957RvMP...29..334G. doi:10.1103/RevModPhys.29.334.
17. Gupta, S. N. (1962). "Quantum Theory of Gravitation". Recent Developments in General Relativity. Pergamon Press. pp. 251–258.
18. Deser, S. (1970). "Self-Interaction and Gauge Invariance". General Relativity and Gravitation. 1: 9–18. arXiv:gr-qc/0411023 . Bibcode:1970GReGr...1....9D. doi:10.1007/BF00759198.
19. Charles Ginenthal. "Newton, Einstein, and Velikovsky".
20. Ohta, Tadayuki; Mann, Robert (1996). "Canonical reduction of two-dimensional gravity for particle dynamics". Classical and Quantum Gravity. 13 (9): 2585–2602. arXiv:gr-qc/9605004 . Bibcode:1996CQGra..13.2585O. doi:10.1088/0264-9381/13/9/022.
21. Sikkema, A E; Mann, R B (1991). "Gravitation and cosmology in (1+1) dimensions". Classical and Quantum Gravity. 8: 219–235. Bibcode:1991CQGra...8..219S. doi:10.1088/0264-9381/8/1/022.
22. Farrugia; Mann; Scott (2007). "N-body Gravity and the Schroedinger Equation". Classical and Quantum Gravity. 24 (18): 4647–4659. arXiv:gr-qc/0611144 . Bibcode:2007CQGra..24.4647F. doi:10.1088/0264-9381/24/18/006.
23. Scott, T.C.; Zhang, Xiangdong; Mann, Robert; Fee, G.J. (2016). "Canonical reduction for dilatonic gravity in 3 + 1 dimensions". Physical Review D. 93 (8): 084017. arXiv:1605.03431 . Bibcode:2016PhRvD..93h4017S. doi:10.1103/PhysRevD.93.084017.
24. Mann, R B; Ohta, T (1997). "Exact solution for the metric and the motion of two bodies in (1+1)-dimensional gravity". Phys. Rev. D. 55 (8): 4723–4747. arXiv:gr-qc/9611008 . Bibcode:1997PhRvD..55.4723M. doi:10.1103/PhysRevD.55.4723.
25. Bellazzini, B.; Csaki, C.; Hubisz, J.; Serra, J.; Terning, J. (2013). "A higgs-like dilaton". Eur. Phys. J. C. 73 (2): 2333. doi:10.1140/epjc/s10052-013-2333-x.
26. Feynman, R. P.; Morinigo, F. B.; Wagner, W. G.; Hatfield, B. (1995). Feynman

lectures on gravitation. Addison-Wesley. ISBN 0-201-62734-5.

27. Hamber, H. W. (2009). Quantum Gravitation - The Feynman Path Integral Approach. Springer Publishing. ISBN 978-3-540-85292-6.

28. Smolin, Lee (2001). Three Roads to Quantum Gravity. Basic Books. pp. 20–25. ISBN 0-465-07835-4. Pages 220–226 are annotated references and guide for further reading.

29. Sakurai, J. J.; Napolitano, Jim J. (2010-07-14). Modern Quantum Mechanics (2 ed.). Pearson. p. 68. ISBN 9780805382914.

30. Novello, Mario; Bergliaffa, Santiago E. (2003-06-11). Cosmology and Gravitation: Xth Brazilian School of Cosmology and Gravitation; 25th Anniversary (1977-2002), Mangaratiba, Rio de Janeiro, Brazil,. Springer Science & Business Media. p. 95. ISBN 9780735401310.

31. A timeline and overview can be found in Rovelli, Carlo (2000). "Notes for a brief history of quantum gravity". arXiv:gr-qc/0006061 . (verify against ISBN 9789812777386)

32. Ashtekar, Abhay (2007). "Loop Quantum Gravity: Four Recent Advances and a Dozen Frequently Asked Questions". 11th Marcel Grossmann Meeting on Recent Developments in Theoretical and Experimental General Relativity. p. 126. arXiv:0705.2222 . Bibcode:2008mgm..conf..126A. doi:10.1142/9789812834300_0008.

33. Schwarz, John H. (2007). "String Theory: Progress and Problems". Progress of Theoretical Physics Supplement. 170: 214–226. arXiv:hep-th/0702219 . Bibcode:2007PThPS.170..214S. doi:10.1143/PTPS.170.214.

34. Donoghue, John F. (editor) (1995). "Introduction to the Effective Field Theory Description of Gravity". In Cornet, Fernando. Effective Theories: Proceedings of the Advanced School, Almunecar, Spain, 26 June–1 July 1995. Singapore: World Scientific. arXiv:gr-qc/9512024 . ISBN 981-02-2908-9.

35. Weinberg, Steven (1996). "Chapters 17–18". The Quantum Theory of Fields II: Modern Applications. Cambridge University Press. ISBN 0-521-55002-5.

36. Goroff, Marc H.; Sagnotti, Augusto; Sagnotti, Augusto (1985). "Quantum gravity at two loops". Physics Letters B. 160: 81–86. Bibcode:1985PhLB..160...81G. doi:10.1016/0370-2693(85)91470-4.

37. An accessible introduction at the undergraduate level can be found in Zwiebach, Barton (2004). A First Course in String Theory. Cambridge University Press. ISBN 0-521-83143-1., and more complete overviews in Polchinski, Joseph (1998). String Theory Vol. I: An Introduction to the Bosonic String. Cambridge University Press. ISBN 0-521-63303-6. and Polchinski, Joseph (1998b). String Theory Vol. II: Superstring Theory and Beyond. Cambridge University Press. ISBN 0-521-63304-4.

38. Ibanez, L. E. (2000). "The second string (phenomenology) revolution". Classical & Quantum Gravity. 17 (5): 1117–1128. arXiv:hep-ph/9911499 . Bibcode:2000CQGra..17.1117I. doi:10.1088/0264-9381/17/5/321.

39. For the graviton as part of the string spectrum, e.g. Green, Schwarz & Witten 1987, sec. 2.3 and 5.3; for the extra dimensions, ibid sec. 4.2.
40. Weinberg, Steven (2000). "Chapter 31". The Quantum Theory of Fields II: Modern Applications. Cambridge University Press. ISBN 0-521-55002-5.
41. Townsend, Paul K. (1996). Four Lectures on M-Theory. ICTP Series in Theoretical Physics. p. 385. arXiv:hep-th/9612121 . Bibcode:1997hepcbconf..385T.
42. Duff, Michael (1996). "M-Theory (the Theory Formerly Known as Strings)". International Journal of Modern Physics A. 11 (32): 5623–5642. arXiv:hep-th/9608117 . Bibcode:1996IJMPA..11.5623D. doi:10.1142/S0217751X96002583.
43. Ashtekar, Abhay (1986). "New variables for classical and quantum gravity". Physical Review Letters. 57 (18): 2244–2247. Bibcode:1986PhRvL..57.2244A. doi:10.1103/PhysRevLett.57.2244. PMID 10033673.
44. Ashtekar, Abhay (1987). "New Hamiltonian formulation of general relativity". Physical Review D. 36 (6): 1587–1602. Bibcode:1987PhRvD..36.1587A. doi:10.1103/PhysRevD.36.1587.
45. Thiemann, Thomas (2006). "Loop Quantum Gravity: An Inside View". Approaches to Fundamental Physics. Lecture Notes in Physics. 721: 185–263. arXiv:hep-th/0608210 . Bibcode:2007LNP...721..185T. doi:10.1007/978-3-540-71117-9_10. ISBN 978-3-540-71115-5.
46. Rovelli, Carlo (1998). "Loop Quantum Gravity". Living Reviews in Relativity. 1. Retrieved 2008-03-13.
47. Ashtekar, Abhay; Lewandowski, Jerzy (2004). "Background Independent Quantum Gravity: A Status Report". Classical & Quantum Gravity. 21 (15): R53–R152. arXiv:gr-qc/0404018 . Bibcode:2004CQGra..21R..53A. doi:10.1088/0264-9381/21/15/R01.
48. Thiemann, Thomas (2003). "Lectures on Loop Quantum Gravity". Lecture Notes in Physics. Lecture Notes in Physics. 631: 41–135. arXiv:gr-qc/0210094 . Bibcode:2003LNP...631...41T. doi:10.1007/978-3-540-45230-0_3. ISBN 978-3-540-40810-9.
49. Rovelli, Carlo (2004). Quantum Gravity. Cambridge University Press. ISBN 0521715962.
50. Isham, Christopher J. (1994). "Prima facie questions in quantum gravity". In Ehlers, Jürgen; Friedrich, Helmut. Canonical Gravity: From Classical to Quantum. Springer. arXiv:gr-qc/9310031 . ISBN 3-540-58339-4.
51. Sorkin, Rafael D. (1997). "Forks in the Road, on the Way to Quantum Gravity". International Journal of Theoretical Physics. 36 (12): 2759–2781. arXiv:gr-qc/9706002 . Bibcode:1997IJTP...36.2759S. doi:10.1007/BF02435709.
52. Loll, Renate (1998). "Discrete Approaches to Quantum Gravity in Four Dimensions". Living Reviews in Relativity. 1: 13. arXiv:gr-qc/9805049 . Bibcode:1998LRR.....1...13L. doi:10.12942/lrr-1998-13. Retrieved 2008-03-09.

53. Sorkin, Rafael D. (2005). "Causal Sets: Discrete Gravity". In Gomberoff, Andres; Marolf, Donald. Lectures on Quantum Gravity. Springer. arXiv:gr-qc/0309009 . ISBN 0-387-23995-2.
54. See Daniele Oriti and references therein.
55. Hawking, Stephen W. (1987). "Quantum cosmology". In Hawking, Stephen W.; Israel, Werner. 300 Years of Gravitation. Cambridge University Press. pp. 631–651. ISBN 0-521-37976-8.
56. Wen 2006
57. See ch. 33 in Penrose 2004 and references therein.
58. Aastrup, J. and Grimstrup, J. M. (27 Apr 2015). "Quantum Holonomy Theory". arXiv:1504.07100 .
59. Hossenfelder, Sabine (2011). "Experimental Search for Quantum Gravity". In V. R. Frignanni. Classical and Quantum Gravity: Theory, Analysis and Applications. Chapter 5: Nova Publishers. ISBN 978-1-61122-957-8.
60. Hossenfelder, Sabine (2010-10-17). V. R. Frignanni, ed. "Experimental Search for Quantum Gravity Chapter 5". Classical and Quantum Gravity: Theory, Analysis and Applications. Nova Publishers. 5 (2011): arXiv:1010.3420. arXiv:1010.3420 . Bibcode:2010arXiv1010.3420H.
61. Camille Carlisle. "First Direct Evidence of Big Bang Inflation". SkyandTelescope.com. Retrieved March 18, 2014.

CHAPTER 8

Relativistic Quantum Mechanics

Chapter Contents

There is no "theory of relativity" -- only two theories that both occasionally fall under the same designator -- the theory of special relativity and the theory of general relativityBoth were devised by the famous physicist Albert Einstein between 1905 and 1915Both these theories upended Newton's classic mechanics, which had held for centuries prior.

INTRODUCTION

In physics, relativistic quantum mechanics (RQM) is any Poincaré covariant formulation of quantum mechanics (QM)This theory is applicable to massive particles propagating at all velocities up to those comparable to the speed of light c, and can accommodate massless particlesThe theory has application in high energy physics,particle physics and accelerator physics, as well as atomic physics, chemistry and condensed matter physicsNon-relativistic quantum mechanics refers to the mathematical formulation of quantum mechanics applied in the context of Galilean relativity, more specifically quantizing the equations of classical mechanics by replacing dynamical variables by operatorsRelativistic quantum mechanics (RQM) is quantum mechanics applied with special relativity, but not general relativityAlthough the earlier formulations, like the Schrödinger picture and Heisenberg picture were originally formulated in a non-relativistic background, these pictures of quantum mechanics also apply with special relativity.

The relativistic formulation is more successful than the original quantum mechanics in some contexts, in particular: the prediction of antimatter, electron spin, spin magnetic moments of elementary spin-1/2 fermions, fine structure, and quantum dynamics of charged particles in electromagnetic fieldsThe key result is the Dirac equation, from which these predictions emerge automaticallyBy contrast, in quantum mechanics, terms have to be introduced artificially into the Hamiltonian operator to achieve agreement with experimental observations.

Nevertheless, RQM is only an approximation to a fully self-consistent relativistic theory of known particle interactions because it does not describe cases where the number of particles changes; for example in matter creation and annihilationBy yet another theoretical advance, a more accurate theory that allows for these occurrences and other predictions is relativistic quantum field theory in which particles are interpreted as field quanta (see article for details).

In this article, the equations are written in familiar 3d vector calculus notation and use hats for operators (not necessarily in the literature), and where space and time components can be collected, tensor index notation is shown also (frequently used in the literature), in addition the Einstein summation convention is usedSI units are used here; Gaussian units and natural units are common alternativesAll equations are in the position representation; for the momentum representation the equations have to be Fourier transformed – see position and momentum space.

COMBINING SPECIAL RELATIVITY AND QUANTUM MECHANICS

One approach is to modify the Schrödinger picture to be consistent with special relativity.

A postulate of quantum mechanics is that the time evolution of any quantum system is given by the Schrödinger equation:

$$i\hbar\frac{\partial}{\partial t}\psi = \hat{H}\psi$$

using a suitable Hamiltonian operator $\hat{H}$ corresponding to the systemThe solution is a complex-valued wavefunction $\psi(\mathbf{r}, t)$, a function of the 3d position vector $\mathbf{r}$ of the particle at time t, describing the behavior of the system.

Every particle has a non-negative spin quantum number sThe number 2s is an integer, odd for fermions and even for bosonsEach s has 2s + 1 z-projection quantum numbers; $\sigma = s, s - 1, .., -s + 1, -s$This is an additional discrete variable the wavefunction requires; $\psi(\mathbf{r}, t, \sigma)$.

Historically, in the early 1920s Pauli, Kronig, Uhlenbeck and Goudsmit were the first to propose the concept of spinThe inclusion of spin in the wavefunction incorporates the Pauli exclusion principle (1925) and the more general spin-statistics theorem (1939) due to Fierz, rederived by Pauli a year laterThis is the explanation for a diverse range of subatomic particle behavior and phenomena: from the electronic configurations of atoms, nuclei (and therefore all elements on the periodic table and their chemistry), to the quark configurations and color charge (hence the properties of baryons and mesons).

A fundamental prediction of special relativity is the relativistic energy–momentum relation; for a particle of rest mass m, and in a particular frame of reference with energy E and 3-momentum **p** with magnitude in terms of the dot product $p = \sqrt{\mathbf{p} \cdot \mathbf{p}}$, it is:

$$E^2 = c^2\mathbf{p}\cdot\mathbf{p} + (mc^2)^2 .$$

These equations are used together with the energy and momentum operators, which are respectively:

$$\hat{E} = i\hbar\frac{\partial}{\partial t}, \quad \hat{\mathbf{p}} = -i\hbar\nabla ,$$

to construct a relativistic wave equation (RWE): a partial differential equation consistent with the energy–momentum relation, and is solved for ψ to predict the quantum dynamics of the particleFor space and time to be placed on equal footing, as in relativity, the orders of space and time partial derivatives should be equal, and ideally as low as possible, so that no initial values of the derivatives need to be specifiedThis is important for probability interpretations, exemplified belowThe lowest possible order of any differential equation is the first (zeroth order derivatives would not form a differential equation).

The Heisenberg picture is another formulation of QM, in which case the wavefunction ψ is time-independent, and the operators A(t) contain the time dependence, governed by the equation of motion:

$$\frac{d}{dt}A = \frac{1}{i\hbar}[A, \hat{H}] + \frac{\partial}{\partial t}A ,$$

This equation is also true in RQM, provided the Heisenberg operators are modified to be consistent with SR.

Historically, around 1926, Schrödinger and Heisenberg show that wave mechanics and matrix mechanics are equivalent, later furthered by Dirac using transformation theory.

A more modern approach to RWEs, first introduced during the time RWEs were developing for particles of any spin, is to apply representations of the Lorentz group.

Space and Time

In classical mechanics and non-relativistic QM, time is an absolute quantity all observers and particles can always agree on, "ticking away" in the background independent of spaceThus in non-relativistic QM one has for a many particle system$\psi(\mathbf{r}_1, \mathbf{r}_2, \mathbf{r}_3, ..., t, \sigma_1, \sigma_2, \sigma_3...)$.

In relativistic mechanics, the spatial coordinates and coordinate time are not absolute; any two observers moving relative to each other can measure different locations and times of eventsThe position and time coordinates combine naturally into a four-dimensional spacetime position **X** = (ct, **r**) corresponding to events, and the energy and 3-momentum combine naturally into the four momentum **P** = (E/c, **p**) of a dynamic particle, as measured in some reference frame, change according to a Lorentz transformation as one measures in a different frame boosted and/or rotated relative the original frame in considerationThe derivative operators, and hence the energy and 3-momentum operators, are also non-invariant and change under Lorentz transformations.

Under a proper orthochronous Lorentz transformation (**r**, t) → Λ(**r**, t) in Minkowski space, all one-particle quantum states ψ_σ locally transform under some representation D of the Lorentz group:

$$\psi_\sigma(\mathbf{r},t) \to D(\Lambda)\psi_\sigma(\Lambda^{-1}(\mathbf{r},t))$$

where D(Λ) is a finite-dimensional representation, in other words a (2s + 1)×(2s + 1) square matrix Again, ψ is thought of as a column vector containing components with the (2s + 1) allowed values of σThe quantum numbers s and σ as well as other labels, continuous or discrete, representing other quantum numbers are suppressedOne value of σ may occur more than once depending on the representation.

Non-Relativistic and Relativistic Hamiltonians

The classical Hamiltonian for a particle in a potential is the kinetic energy **p·p**/2m plus the potential energy V(**r**, t), with the corresponding quantum operator in the Schrödinger picture:

$$\hat{H} = \frac{\hat{\mathbf{p}}\cdot\hat{\mathbf{p}}}{2m} + V(\mathbf{r},t)$$

and substituting this into the above Schrödinger equation gives a non-relativistic QM equation for the wavefunction: the procedure is a straightforward substitution of a simple expressionBy contrast this is not as easy in RQM; the energy–momentum equation is quadratic in energy and momentum leading to difficultiesNaively setting:

$$\hat{H} = \hat{E} = \sqrt{c^2\hat{\mathbf{p}}\cdot\hat{\mathbf{p}} + (mc^2)^2} \quad \Rightarrow \quad i\hbar\frac{\partial}{\partial t}\psi = \sqrt{c^2\hat{\mathbf{p}}\cdot\hat{\mathbf{p}} + (mc^2)^2}\,\psi$$

is not helpful for several reasonsThe square root of the operators cannot be used as it stands; it would have to be expanded in a power series before the momentum operator, raised to a power in each term, could act on ψAs a result of the power series, the space and time derivatives are completely asymmetric: infinite-order in space derivatives but only first order in the time derivative, which is inelegant and unwieldyAgain, there is the problem of the non-invariance of the energy operator, equated to the square root which is also not invariantAnother problem, less obvious and more severe, is that it can be shown to be nonlocal and can even violate causality: if the particle is initially localized at a point $\mathbf{r}_0$ so that $\psi(\mathbf{r}_0, t = 0)$ is finite and zero elsewhere, then at any later time the equation predicts delocalization ψ(**r**, t) ≠ 0 everywhere, even for |**r**| > ct which means the particle could arrive at a point before a pulse of light couldThis would have to be remedied by the additional constraint ψ(|**r**| > ct, t) = 0.

There is also the problem of incorporating spin in the Hamiltonian, which isn't a prediction of the non-relativistic Schrödinger theoryParticles with spin have a corresponding spin magnetic moment quantized in units of μ_B, the Bohr magneton:

$$\hat{\boldsymbol{\mu}}_S = -\frac{g\mu_B}{\hbar}\hat{\mathbf{S}}, \quad |\boldsymbol{\mu}_S| = -g\mu_B\sigma,$$

where g is the (spin) g-factor for the particle, and **S** the spin operator, so they interact with electromagnetic fieldsFor a particle in an externally applied magnetic field **B**, the interaction term

$$\hat{H}_B = -\mathbf{B} \cdot \hat{\boldsymbol{\mu}}_S$$

has to be added to the above non-relativistic HamiltonianOn the contrary; a relativistic Hamiltonian introduces spin automatically as a requirement of enforcing the relativistic energy-momentum relation.

Relativistic Hamiltonians are analogous to those of non-relativistic QM in the following respect; there are terms including rest mass and interaction terms with externally applied fields, similar to the classical potential energy term, as well as momentum terms like the classical kinetic energy termA key difference is that relativistic Hamiltonians contain spin operators in the form of matrices, in which the matrix multiplication runs over the spin index σ, so in general a relativistic Hamiltonian:

$$\hat{H} = \hat{H}(\mathbf{r}, t, \hat{\mathbf{p}}, \hat{\mathbf{S}})$$

is a function of space, time, and the momentum and spin operators.

The Klein–Gordon and Dirac Equations for Free Particles

Substituting the energy and momentum operators directly into the energy–momentum relation may at first sight seem appealing, to obtain the Klein–Gordon equation:

$$\hat{E}^2\psi = c^2\hat{\mathbf{p}} \cdot \hat{\mathbf{p}}\psi + (mc^2)^2\psi,$$

and was discovered by many people because of the straightforward way of obtaining it, notably by Schrödinger in 1925 before he found the non-relativistic equation named after him, and by Klein and Gordon in 1927, who included electromagnetic interactions in the equationThis is relativistically invariant, yet this equation alone isn›t a sufficient foundation for RQM for a few reasons; one is that negative-energy states are solutions, another is the density (given below), and this equation as it stands is only applicable to spinless particlesThis equation can be factored into the form:

$$\left(\hat{E} - c\boldsymbol{\alpha} \cdot \hat{\mathbf{p}} - \beta mc^2\right)\left(\hat{E} + c\boldsymbol{\alpha} \cdot \hat{\mathbf{p}} + \beta mc^2\right)\psi = 0,$$

where $\boldsymbol{\alpha} = (\alpha_1, \alpha_2, \alpha_3)$ and β are not simply numbers or vectors, but 4 × 4 Hermitian matrices that are required to anticommute for i ≠ j:

$$\alpha_i\beta = -\beta\alpha_i, \quad \alpha_i\alpha_j = -\alpha_j\alpha_i,$$

and square to the identity matrix:

$$\alpha_i^2 = \beta^2 = I,$$

so that terms with mixed second-order derivatives cancel while the second-order derivatives purely in space and time remain.

The first factor:

$$\left(\hat{E} - c\boldsymbol{\alpha} \cdot \hat{\mathbf{p}} - \beta mc^2\right)\psi = 0 \quad \Leftrightarrow \quad \hat{H} = c\boldsymbol{\alpha} \cdot \hat{\mathbf{p}} + \beta mc^2$$

is the Dirac equationThe other factor is also the Dirac equation, but for a particle of negative massEach factor is relativistically invariantThe reasoning can be done the other way round: propose the Hamiltonian in the above form, as Dirac did in 1928, then pre-multiply the equation by the other factor of operators E + c**α** · **p** + βmc^2, and comparison with the KG equation determines the constraints on **α** and βThe positive mass equation can continue to be used without loss of continuityThe matrices multiplying ψ suggest it isn›t a scalar wavefunction as permitted in the KG equation, but must instead be a four-component entityThe Dirac equation still predicts negative energy solutions, so Dirac postulated that negative energy states are always occupied, because according to the Pauli principle, electronic transitions from positive to negative energy levels in atoms would be forbiddenSee Dirac sea for details.

Densities and Currents

In non-relativistic quantum mechanics, the square-modulus of the wavefunction ψ gives the probability density functionρ = |ψ|^{2}This is the Copenhagen interpretation, circa 1927In RQM, while ψ(**r**, t) is a wavefunction, the probability interpretation is not the same as in non-relativistic QMSome RWEs do not predict a probability density ρ or probability current**j** (really meaning probability current density) because they are not positive definite functions of space and timeThe Dirac equation does:

$$\rho = \psi^\dagger \psi, \quad \mathbf{j} = \psi^\dagger \gamma^0 \boldsymbol{\gamma} \psi \quad \rightleftharpoons \quad J^\mu = \psi^\dagger \gamma^0 \gamma^\mu \psi$$

where the dagger denotes the Hermitian adjoint (authors usually write ψ = ψ†γ0 for the Dirac adjoint and Jμ is the probability four-current, while the Klein–Gordon equation does not:

$$\rho = \frac{i\hbar}{2mc^2}\left(\psi^* \frac{\partial \psi}{\partial t} - \psi \frac{\partial \psi^*}{\partial t}\right), \quad \mathbf{j} = -\frac{i\hbar}{2m}\left(\psi^* \nabla \psi - \psi \nabla \psi^*\right) \quad \rightleftharpoons \quad J^\mu = \frac{i\hbar}{2m}\left(\psi^* \partial^\mu \psi - \psi \partial^\mu \psi^*\right)$$

where ∂μ is the four gradient. Since the initial values of both ψ and ∂ψ/∂t may be freely chosen, the density can be negative

Instead, what appears look at first sight a "probability density" and "probability current" has to be reinterpreted as charge density and current density when multiplied by electric chargeThen, the wavefunction ψ is not a wavefunction at all, but reinterpreted as a fieldThe density and current of electric charge always satisfy a continuity equation:

$$\frac{\partial \rho}{\partial t} + \nabla \cdot \mathbf{J} = 0 \quad \rightleftharpoons \quad \partial_\mu J^\mu = 0,$$

as charge is a conserved quantityProbability density and current also satisfy a continuity equation because probability is conserved, however this is only possible in the absence of interactions.

SPIN AND ELECTROMAGNETICALLY INTERACTING PARTICLES

Including interactions in RWEs is generally difficultMinimal coupling is a simple way to include the electromagnetic interactionFor one charged particle of electric charge q in an electromagnetic field, given by the magnetic vector potential **A**(**r**, t) defined by the magnetic field $\mathbf{B} = \nabla \times \mathbf{A}$, and electric scalar potential $\phi(\mathbf{r}, t)$, this is:

$$\hat{E} \to \hat{E} - q\phi, \quad \hat{\mathbf{p}} \to \hat{\mathbf{p}} - q\mathbf{A} \quad \rightleftharpoons \quad \hat{P}_\mu \to \hat{P}_\mu - qA_\mu$$

where $P\mu$ is the four-momentum that has a corresponding 4-momentum operator, and $A\mu$ the four-potential. In the following, the non-relativistic limit refers to the limiting cases:

$$E - e\phi \approx mc^2, \quad \mathbf{p} \approx m\mathbf{v},$$

that is, the total energy of the particle is approximately the rest energy for small electric potentials, and the momentum is approximately the classical momentum.

Spin-0

In RQM, the KG equation admits the minimal coupling prescription;

$$(\hat{E} - q\phi)^2 \psi = c^2 (\hat{\mathbf{p}} - q\mathbf{A})^2 \psi + (mc^2)^2 \psi \quad \rightleftharpoons \quad \left[(\hat{P}_\mu - qA_\mu)(\hat{P}^\mu - qA^\mu) - (mc)^2\right] \psi = 0.$$

In the case where the charge is zero, the equation reduces trivially to the free KG equation so nonzero charge is assumed belowThis is a scalar equation that is invariant under the irreducible one-dimensional scalar (0,0) representation of the Lorentz groupThis means that all of its solutions will belong to a direct sum of (0,0) representationsSolutions that do not belong to the irreducible (0,0) representation will have two or more independent componentsSuch solutions cannot in general describe particles with nonzero spin since spin components are not independentOther constraint will have to be imposed for that, e.gthe Dirac equation for spin 1/2, see belowThus if a system satisfies the KG equation only, it can only be interpreted as a system with zero spin.

The electromagnetic field is treated classically according to Maxwell's equations and the particle is described by a wavefunction, the solution to the KG equationThe equation is, as it stands, not always very useful, because massive spinless particles, such as the π-mesons, experience the much stronger strong interaction in addition to the electromagnetic interactionIt does, however, correctly describe charged spinless bosons in the absence of other interactions.

The KG equation is applicable to spinless charged bosons in an external electromagnetic potentialAs such, the equation cannot be applied to the description of atoms, since the electron is a spin 1/2 particleIn the non-relativistic limit the equation reduces to the Schrödinger equation for a spinless charged particle in an electromagnetic field:

$$\left(i\hbar\frac{\partial}{\partial t} - q\phi\right)\psi = \frac{1}{2m}(\hat{\mathbf{p}} - q\mathbf{A})^2\psi \quad \Leftrightarrow \quad \hat{H} = \frac{1}{2m}(\hat{\mathbf{p}} - q\mathbf{A})^2 + q\phi.$$

Spin-1/2

Non relativistically, spin was phenomenologically introduced in the Pauli equation by Pauli in 1927 for particles in an electromagnetic field:

$$\left(i\hbar\frac{\partial}{\partial t} - q\phi\right)\psi = \left[\frac{1}{2m}(\boldsymbol{\sigma}\cdot(\mathbf{p} - q\mathbf{A}))^2\right]\psi \quad \Leftrightarrow \quad \hat{H} = \frac{1}{2m}(\boldsymbol{\sigma}\cdot(\mathbf{p} - q\mathbf{A}))^2 + q\phi$$

by means of the 2 × 2 Pauli matrices, and ψ is not just a scalar wavefunction as in the non-relativistic Schrödinger equation, but a two-component spinor field:

$$\psi = \begin{pmatrix}\psi_\uparrow \\ \psi_\downarrow\end{pmatrix}$$

where the subscripts ↑ and ↓ refer to the "spin up" (σ = +1/2) and "spin down" (σ = −1/2) states. In RQM, the Dirac equation can also incorporate minimal coupling, rewritten from above;

$$\left(i\hbar\frac{\partial}{\partial t} - q\phi\right)\psi = \gamma^0\left[c\boldsymbol{\gamma}\cdot(\hat{\mathbf{p}} - q\mathbf{A}) - mc^2\right]\psi \quad \rightleftharpoons \quad \left[\gamma^\mu(\hat{P}_\mu - qA_\mu) - mc^2\right]\psi = 0$$

and was the first equation to accurately predict spin, a consequence of the 4 × 4 gamma matrices$\gamma^0 = \beta$, $\boldsymbol{\gamma} = (\gamma_1, \gamma_2, \gamma_3) = \beta\boldsymbol{\alpha} = (\beta\alpha_1, \beta\alpha_2, \beta\alpha_3)$There is a 4 × 4 identity matrix pre-multiplying the energy operator (including the potential energy term), conventionally not written for simplicity and clarity (i.etreated like the number 1)Here ψ is a four-component spinor field, which is conventionally split into two two-component spinors in the form:

$$\psi = \begin{pmatrix}\psi_+ \\ \psi_-\end{pmatrix} = \begin{pmatrix}\psi_{+\uparrow} \\ \psi_{+\downarrow} \\ \psi_{-\uparrow} \\ \psi_{-\downarrow}\end{pmatrix}$$

The 2-spinor ψ_+ corresponds to a particle with 4-momentum (E, **p**) and charge q and two spin states (σ = ±1/2, as before)The other 2-spinor ψ_- corresponds to a similar particle with the same mass and spin states, but negative 4-momentum −(E, **p**)and negative charge −q, that is, negative energy states, time-reversed momentum, and negated chargeThis was the first interpretation and prediction of a particle and corresponding antiparticleSee Dirac spinor and bispinor for further description of these spinorsIn the non-relativistic limit the Dirac equation reduces to the Pauli equation (see Dirac equation for how)When applied a one-electron atom or ion, setting **A** = **0** and ϕ to the appropriate electrostatic potential, additional relativistic terms include the spin-orbit interaction, electron gyromagnetic ratio, and Darwin termIn ordinary QM these terms have to be put in by hand and treated using perturbation theoryThe positive energies do account accurately for the fine structure.

Within RQM, for massless particles the Dirac equation reduces to:

$$\left(\frac{\hat{E}}{c} + \boldsymbol{\sigma}\cdot\hat{\mathbf{p}}\right)\psi_+ = 0, \quad \left(\frac{\hat{E}}{c} - \boldsymbol{\sigma}\cdot\hat{\mathbf{p}}\right)\psi_- = 0 \quad \rightleftharpoons \quad \sigma^\mu\hat{P}_\mu\psi_+ = 0, \quad \sigma_\mu\hat{P}^\mu\psi_- = 0,$$

the first of which is the Weyl equation, a considerable simplification applicable for massless neutrinosThis time there is a 2 × 2 identity matrix pre-multiplying the energy operator conventionally not writtenIn RQM it is useful to take this as the zeroth Pauli matrix σ_0 which couples to the energy operator (time derivative), just as the other three matrices couple to the momentum operator (spatial derivatives).

The Pauli and gamma matrices were introduced here, in theoretical physics, rather than pure mathematics itselfThey have applications to quaternions and to the SO(2) and SO(3) Lie groups, because they satisfy the important commutator and anticommutator relations respectively:

$$[\sigma_a, \sigma_b] = 2i\varepsilon_{abc}\sigma_c\,, \quad [\sigma_a, \sigma_b]_+ = 2\delta_{ab}\sigma_0$$

where εabc is the three-dimensional Levi-Civita symbol. The gamma matrices form bases in Clifford algebra, and have a connection to the components of the flat spacetime Minkowski metric ηαβ in the anticommutation relation:

$$[\gamma^\alpha, \gamma^\beta]_+ = \gamma^\alpha\gamma^\beta + \gamma^\beta\gamma^\alpha = 2\eta^{\alpha\beta}\,,$$

(This can be extended to curved spacetime by introducing vierbeins, but is not the subject of special relativity).

In 1929, the Breit equation was found to describe two or more electromagnetically interacting massive spin-1/2 fermions to first-order relativistic corrections; one of the first attempts to describe such a relativistic quantum many-particle systemThis is, however, still only an approximation, and the Hamiltonian includes numerous long and complicated sums.

Helicity and Chirality

The helicity operator is defined by;

$$\hat{h} = \hat{\mathbf{S}} \cdot \frac{\hat{\mathbf{p}}}{|\mathbf{p}|} = \hat{\mathbf{S}} \cdot \frac{c\hat{\mathbf{p}}}{\sqrt{E^2 - (m_0c^2)^2}}$$

where **p** is the momentum operator, **S** the spin operator for a particle of spin s, E is the total energy of the particle, and m_0 its rest massHelicity indicates the orientations of the spin and translational momentum vectorsHelicity is frame-dependent because of the 3-momentum in the definition, and is quantized due to spin quantization, which has discrete positive values for parallel alignment, and negative values for antiparallel alignment.

A automatic occurrence in the Dirac equation (and the Weyl equation) is the projection of the spin-1/2 operator on the 3-momentum (times c), **σ** · c **p**, which is the helicity (for the spin-1/2 case) times $\sqrt{E^2 - (m_0c^2)^2}$.

For massless particles the helicity simplifies to:

$$\hat{h} = \hat{\mathbf{S}} \cdot \frac{c\hat{\mathbf{p}}}{E}$$

Higher Spins

The Dirac equation can only describe particles of spin-1/2Beyond the Dirac equation, RWEs have been applied to free particles of various spinsIn 1936, Dirac extended his equation to all fermions, three years later Fierz and Pauli rederived the same equationThe Bargmann–Wigner equations were found in 1948 using Lorentz group theory, applicable for all free particles with any spinConsidering the factorization of the KG equation above, and more rigorously by Lorentz group theory, it becomes apparent to introduce spin in the form of matrices.

The wavefunctions are multicomponent spinor fields, which can be represented as column vectors of functions of space and time:

$$\psi(\mathbf{r},t) = \begin{bmatrix} \psi_{\sigma=s}(\mathbf{r},t) \\ \psi_{\sigma=s-1}(\mathbf{r},t) \\ \vdots \\ \psi_{\sigma=-s+1}(\mathbf{r},t) \\ \psi_{\sigma=-s}(\mathbf{r},t) \end{bmatrix} \rightleftharpoons \psi(\mathbf{r},t)^{\dagger} = \left[\psi_{\sigma=s}(\mathbf{r},t)^{\star} \quad \psi_{\sigma=s-1}(\mathbf{r},t)^{\star} \quad \cdots \quad \psi_{\sigma=-s+1}(\mathbf{r},t)^{\star} \quad \psi_{\sigma=-s}(\mathbf{r},t)^{\star}\right]$$

where the expression on the right is the Hermitian conjugate. For a massive particle of spin s, there are 2s + 1 components for the particle, and another 2s + 1 for the corresponding antiparticle (there are 2s + 1 possible σ values in each case), altogether forming a 2(2s + 1)-component spinor field:

$$\psi(\mathbf{r},t) = \begin{bmatrix} \psi_{+,\sigma=s}(\mathbf{r},t) \\ \psi_{+,\sigma=s-1}(\mathbf{r},t) \\ \vdots \\ \psi_{+,\sigma=-s+1}(\mathbf{r},t) \\ \psi_{+,\sigma=-s}(\mathbf{r},t) \\ \psi_{-,\sigma=s}(\mathbf{r},t) \\ \psi_{-,\sigma=s-1}(\mathbf{r},t) \\ \vdots \\ \psi_{-,\sigma=-s+1}(\mathbf{r},t) \\ \psi_{-,\sigma=-s}(\mathbf{r},t) \end{bmatrix} \rightleftharpoons \psi(\mathbf{r},t)^{\dagger}\left[\psi_{+,\sigma=s}(\mathbf{r},t)^{\star} \quad \psi_{+,\sigma=s-1}(\mathbf{r},t)^{\star} \quad \cdots \quad \psi_{-,\sigma=-s}(\mathbf{r},t)^{\star}\right]$$

with the + subscript indicating the particle and − subscript for the antiparticleHowever, for massless particles of spin s, there are only ever twocomponent spinor fields; one is for the particle in one helicity state corresponding to +s and the other for the antiparticle in the opposite helicity state corresponding to −s:

$$\psi(\mathbf{r},t) = \begin{pmatrix} \psi_{+}(\mathbf{r},t) \\ \psi_{-}(\mathbf{r},t) \end{pmatrix}$$

According to the relativistic energy-momentum relation, all massless particles travel at the speed of light, so particles traveling at the speed of light are also described by two-component spinorsHistorically, Élie Cartan found the most general form of spinors in 1913, prior to the spinors revealed in the RWEs following the year 1927.

For equations describing higher-spin particles, the inclusion of interactions is nowhere near as simple minimal coupling, they lead to incorrect predictions and self-inconsistenciesFor spin greater than ħ/2, the RWE is not fixed by the particle›s mass, spin, and electric charge; the electromagnetic moments (electric dipole moments and magnetic dipole moments) allowed by the spin quantum number are arbitrary(Theoretically, magnetic charge would contribute also)For example, the spin-1/2 case only allows a magnetic dipole, but for spin-1 particles magnetic quadrupoles and electric dipoles are also possibleFor more on this topic, see multipole expansion and (for example) Cédric Lorcé (2009).

VELOCITY OPERATOR

The Schrödinger/Pauli velocity operator can be defined for a massive particle using the classical definition **p** = m **v**, and substituting quantum operators in the usual way:

$$\hat{\mathbf{v}} = \frac{1}{m}\hat{\mathbf{p}}$$

which has eigenvalues that take *any* value. In RQM, the Dirac theory, it is:

$$\hat{\mathbf{v}} = \frac{i}{\hbar}\left[\hat{H}, \hat{\mathbf{r}}\right]$$

which must have eigenvalues between ±cSee Foldy–Wouthuysen transformation for more theoretical background.

Relativistic Quantum Lagrangians

The Hamiltonian operators in the Schrödinger picture are one approach to forming the differential equations for ψAn equivalent alternative is to determine a Lagrangian (really meaning Lagrangian density), then generate the differential equation by the field-theoretic Euler–Lagrange equation:

$$\partial_\mu\left(\frac{\partial\mathcal{L}}{\partial(\partial_\mu\psi)}\right) - \frac{\partial\mathcal{L}}{\partial\psi} = 0$$

For some RWEs, a Lagrangian can be found by inspection. For example, the Dirac Lagrangian is:

$$\mathcal{L} = \overline{\psi}(\gamma^\mu P_\mu - mc)\psi$$

and Klein–Gordon Lagrangian is:

$$\mathcal{L} = -\frac{\hbar^2}{m}\eta^{\mu\nu}\partial_\mu\psi^*\partial_\nu\psi - mc^2\psi^*\psi\,.$$

This is not possible for all RWEs; and is one reason the Lorentz group theoretic approach is important and appealing: fundamental invariance and symmetries in space and time can be used to derive RWEs using appropriate group representationsThe Lagrangian approach with field interpretation of ψ is the subject of QFT rather than RQM: Feynman›s path integral formulation uses invariant Lagrangians rather than Hamiltonian operators, since the latter can become extremely complicated, see (for example) SWeinberg (1995).

RELATIVISTIC QUANTUM ANGULAR MOMENTUM

In non-relativistic QM, the angular momentum operator is formed from the classical pseudovector definition **L** = **r** × **p**In RQM, the position and momentum operators are inserted directly where they appear in the orbital relativistic angular momentum tensor defined from the four-dimensional position and momentum of the particle, equivalently a bivector in the exterior algebra formalism:

$$M^{\alpha\beta} = X^{\alpha}P^{\beta} - X^{\beta}P^{\alpha} = 2X^{[\alpha}P^{\beta]} \quad \rightleftharpoons \quad \mathbf{M} = \mathbf{X} \wedge \mathbf{P},$$

which are six components altogether: three are the non-relativistic 3-orbital angular momenta; M12 = L3, M23 = L1, M31 = L2, and the other three M01, M02, M03 are boosts of the centre of mass of the rotating object. An additional relativistic-quantum term has to be added for particles with spin. For a particle of rest mass m, the total angular momentum tensor is

$$J^{\alpha\beta} = 2X^{[\alpha}P^{\beta]} + \frac{1}{m^2}\varepsilon^{\alpha\beta\gamma\delta}W_{\gamma}p_{\delta} \quad \rightleftharpoons \quad \mathbf{J} = \mathbf{X} \wedge \mathbf{P} + \frac{1}{m^2} \star (\mathbf{W} \wedge \mathbf{P})$$

where the star denotes the Hodge dual, and

$$W_{\alpha} = \frac{1}{2}\varepsilon_{\alpha\beta\gamma\delta}M^{\beta\gamma}p^{\delta} \quad \rightleftharpoons \quad \mathbf{W} = \star(\mathbf{M} \wedge \mathbf{P})$$

is the Pauli–Lubanski pseudovectorFor more on relativistic spin, see (for example) S.MTroshin and N.ETyurin (1994).

Thomas Precession and Spin-Orbit Interactions

In 1926 the Thomas precession is discovered: relativistic corrections to the spin of elementary particles with application in the spin–orbit interaction of atoms and rotation of macroscopic objectsIn 1939 Wigner derived the Thomas precession.

In classical electromagnetism and special relativity, an electron moving with a velocity **v** through an electric field **E** but not a magnetic field **B**, will in its own frame of reference experience a Lorentz-transformed magnetic field **B′**:

$$\mathbf{B}' = \frac{\mathbf{E} \times \mathbf{v}}{c^2\sqrt{1 - (v/c)^2}}.$$

In the non-relativistic limit $v \ll c$:

$$\mathbf{B}' = \frac{\mathbf{E} \times \mathbf{v}}{c^2},$$

so the non-relativistic spin interaction Hamiltonian becomes:

$$\hat{H} = -\mathbf{B}' \cdot \hat{\boldsymbol{\mu}}_S = -\left(\mathbf{B} + \frac{\mathbf{E} \times \mathbf{v}}{c^2}\right) \cdot \hat{\boldsymbol{\mu}}_S,$$

where the first term is already the non-relativistic magnetic moment interaction, and the second term the relativistic correction of order $(v/c)^2$, but this disagrees with experimental atomic spectra by a factor of 1/2It was pointed out by LThomas that there is a second relativistic effect: an electric field component perpendicular to the electron velocity causes an additional acceleration of the electron perpendicular to its instantaneous velocity, so the electron moves in a curved pathThe electron moves in a rotating frame of reference, and this additional precession of the electron is called the Thomas precessionIt can be shown that the net result of this effect is that the spin–orbit interaction is reduced by half, as if the magnetic field experienced by the electron has only one-half the value, and the relativistic correction in the Hamiltonian is:

$$\hat{H} = -\mathbf{B}' \cdot \hat{\boldsymbol{\mu}}_S = -\left(\mathbf{B} + \frac{\mathbf{E} \times \mathbf{v}}{2c^2}\right) \cdot \hat{\boldsymbol{\mu}}_S .$$

In the case of RQM, the factor of 1/2 is predicted by the Dirac equation.

The events which led to and established RQM, and the continuation beyond into quantum electrodynamics (QED), are summarized belowMore than half a century of experimental and theoretical research from the 1890s through to the 1950s in the new and mysterious quantum theory as it was up and coming revealed that a number of phenomena cannot be explained by QM aloneSR, found at the turn of the 20th century, was found to be a necessary component, leading to unification: RQMTheoretical predictions and experiments mainly focused on the newly found atomic physics, nuclear physics, and particle physics; by considering spectroscopy, diffraction and scattering of particles, and the electrons and nuclei within atoms and moleculesNumerous results are attributed to the effects of spin.

Relativistic Description of Particles in Quantum Phenomena

Einstein in 1905 explained of the photoelectric effect; a particle description of light as photonsIn 1916, Sommerfeld explains fine structure; the splitting of the spectral lines of atoms due to first order relativistic correctionsThe Compton effect of 1923 provided more evidence that special relativity does apply; in this case to a particle description of photon–electron scatteringde Broglie extends wave–particle duality to matter: the de Broglie relations, which are consistent with special relativity and quantum mechanicsBy 1927, Davisson and Germer and separately GThomson successfully diffract electrons, providing experimental evidence of wave-particle duality.

Experiments

- 1897 JJThomson discovers the electron and measures its mass-to-charge ratioDiscovery of the Zeeman effect: the splitting a spectral line into several components in the presence of a static magnetic field.
- 1908 Millikan measures the charge on the electron and finds experimental evidence of its quantization, in the oil drop experiment.
- 1911 Alpha particle scattering in the Geiger–Marsden experiment, led by Rutherford, showed that atoms possess an internal structure: the atomic nucleus.
- 1913 The Stark effect is discovered: splitting of spectral lines due to a static electric field (compare with the Zeeman effect).
- 1922 Stern–Gerlach experiment: experimental evidence of spin and its quantization.
- 1924 Stoner studies splitting of energy levels in magnetic fields.
- 1932 Experimental discovery of the neutron by Chadwick, and positrons by

Anderson, confirming the theoretical prediction of positrons.

- 1958 Discovery of the Mössbauer effect: resonant and recoil-free emission and absorption of gamma radiation by atomic nuclei bound in a solid, useful for accurate measurements of gravitational redshift and time dilation, and in the analysis of nuclear electromagnetic moments in hyperfine interactions.

Quantum Non-Locality and Relativistic Locality

In 1935; Einstein, Rosen, Podolsky published a paper concerning quantum entanglement of particles, questioning quantum nonlocality and the apparent violation of causality upheld in SR: particles can appear to interact instantaneously at arbitrary distancesThis was a misconception since information is not and cannot be transferred in the entangled states; rather the information transmission is in the process of measurement by two observers (one observer has to send a signal to the other, which cannot exceed c)QM does not violate SRIn 1959, Bohm and Aharonov publish a paper on the Aharonov–Bohm effect, questioning the status of electromagnetic potentials in QMThe EM field tensor and EM 4-potential formulations are both applicable in SR, but in QM the potentials enter the Hamiltonian (see above) and influence the motion of charged particles even in regions where the fields are zeroIn 1964, Bell›s theorem was published in a paper on the EPR paradox,showing that QM cannot be derived from local hidden variable theories if locality is to be maintained.

The Lamb Shift

In 1947 the Lamb shift was discovered: a small difference in the $^2S_{1/2}$ and $^2P_{1/2}$ levels of hydrogen, due to the interaction between the electron and vacuumLamb and Retherford experimentally measure stimulated radio-frequency transitions the $^2S_{1/2}$ and $^2P_{1/2}$ hydrogen levels by microwave radiationAn explanation of the Lamb shift is presented by BethePapers on the effect were published in the early 1950s.

Development of Quantum Electrodynamics

- 1943 Tomonaga begins work on renormalization, influential in QED.
- 1947 Schwinger calculates the anomalous magnetic moment of the electronKusch measures of the anomalous magnetic electron moment, confirming one of QED›s great predictions.

REFERENCES

1. D.HPerkins (2000)Introduction to High Energy PhysicsCambridge University PressISBN 0-521-62196-8.
2. BRMartin, G.ShawParticle PhysicsManchester Physics Series (3rd ed.)John Wiley & Sonsp3ISBN 978-0-470-03294-7.

3. M.Reiher, A.Wolf (2009)Relativistic Quantum ChemistryJohn Wiley & SonsISBN 3-527-62749-9.
4. PStrange (1998)Relativistic Quantum Mechanics: With Applications in Condensed Matter and Atomic PhysicsCambridge University PressISBN 0-521-56583-9.
5. PMohn (2003)Magnetism in the Solid State: An IntroductionSpringer Series in Solid-State Sciences Series134Springerp6ISBN 3-540-43183-7.
6. BRMartin, G.ShawParticle PhysicsManchester Physics Series (3rd ed.)John Wiley & Sonspp5–6ISBN 978-0-470-03294-7.
7. AMessiah (1981)Quantum Mechanics2North-Holland Publishing Companyp875ISBN 0-7204-0045-7.
8. J.RForshaw; A.GSmith (2009)Dynamics and RelativityManchester Physics SeriesJohn Wiley & Sonspp258–259ISBN 978-0-470-01460-8.
9. WGreiner (2000)Relativistic Quantum MechanicsWave Equations (3rd ed.) Springerp70ISBN 3-540-67457-8.
10. AWachter (2011)"Relativistic quantum mechanics"Springerp34ISBN 90-481-3645-8.
11. Weinberg, S(1964)"Feynman Rules for Any spin" (PDF)PhysRev133 (5B): B1318–B1332Bibcode:1964PhRv..133.1318Wdoi:10.1103/PhysRev.133.B1318.; Weinberg, S(1964)"Feynman Rules for Any spinIIMassless Particles" (PDF) PhysRev134 (4B): B882–B896Bibcode:1964PhRv..134..882Wdoi:10.1103/PhysRev.134.B882.; Weinberg, S(1969)"Feynman Rules for Any spinIII" (PDF)PhysRev181 (5): 1893–1899Bibcode:1969PhRv..181.1893Wdoi:10.1103/PhysRev.181.1893.
12. KMasakatsu (2012)"Superradiance Problem of Bosons and Fermions for Rotating Black Holes in Bargmann–Wigner Formulation"arXiv:1208.0644 .
13. C.BParker (1994)McGraw Hill Encyclopaedia of Physics (2nd ed.)McGraw Hillpp1193–1194ISBN 0-07-051400-3.
14. RResnick; REisberg (1985)Quantum Physics of Atoms, Molecules, Solids, Nuclei and Particles (2nd ed.)John Wiley & Sonsp274ISBN 978-0-471-87373-0.
15. L.DLandau; E.MLifshitz (1981)Quantum Mechanics Non-Relativistic Theory3Elsevierp455ISBN 0-08-050348-9.
16. YPeleg; RPnini; EZaarur; EHecht (2010)Quantum MechanicsShaum's outlines (2nd ed.)McGraw–Hillp181ISBN 978-0-07-162358-2.
17. EAbers (2004)Quantum MechanicsAddison Wesleyp425ISBN 978-0-13-146100-0.
18. AWachter (2011)"Relativistic quantum mechanics"Springerp5ISBN 90-481-3645-8.
19. EAbers (2004)Quantum MechanicsAddison Wesleyp415ISBN 978-0-13-146100-0.
20. RPenrose (2005)The Road to RealityVintage Bookspp620–621ISBN 978-0-09-944068-0.

21. Bransden, BH; Joachain, CJ (1983)Physics of Atoms and Molecules (1st ed.) Prentice Hallp634ISBN 0-582-44401-2.
22. W.TGrandy (1991)Relativistic quantum mechanics of leptons and fieldsSpringerp54ISBN 0-7923-1049-7.
23. EAbers (2004)Quantum MechanicsAddison Wesleyp423ISBN 978-0-13-146100-0.
24. DMcMahon (2008)Quantum Field TheoryDemystifiedMcGraw Hillp114ISBN 978-0-07-154382-8.
25. Bransden, BH; Joachain, CJ (1983)Physics of Atoms and Molecules (1st ed.) Prentice Hallpp632–635ISBN 0-582-44401-2.
26. C.BParker (1994)McGraw Hill Encyclopaedia of Physics (2nd ed.)McGraw Hillp1194ISBN 0-07-051400-3..
27. PLabelle (2010)SupersymmetryDemystifiedMcGraw-HillISBN 978-0-07-163641-4.
28. SEsposito (2011)"Searching for an equation: Dirac, Majorana and the others"arXiv:1110.6878 .
29. Bargmann, V.; Wigner, EP(1948)"Group theoretical discussion of relativistic wave equations" (PDF)ProcNatlAcadSciU.S.A34 (5): 211–23Bibcode:1948PNAS...34..211Bdoi:10.1073/pnas.34.5.211.
30. EWigner (1937)"On Unitary Representations Of The Inhomogeneous Lorentz Group" (PDF)Annals of Mathematics40 (1): 149Bibcode:1939AnMat..40..149Wdoi:10.2307/1968551.
31. TJaroszewicz; P.S Kurzepa (1992)"Geometry of spacetime propagation of spinning particles"Annals of PhysicsBibcode:1992AnPhy.216..226Jdoi:10.1016/0003-4916(92)90176-M.
32. Cédric Lorcé (2009)"Electromagnetic Properties for Arbitrary Spin Particles: Part 1 – Electromagnetic Current and Multipole Decomposition"arXiv:0901.4199 .
33. Cédric Lorcé (2009)"Electromagnetic Properties for Arbitrary Spin Particles: Part 2 – Natural Moments and Transverse Charge Densities"arXiv:0901.4200 .
34. PStrange (1998)Relativistic Quantum Mechanics: With Applications in Condensed Matter and Atomic PhysicsCambridge University Pressp206ISBN 0-521-56583-9.
35. PLabelle (2010)SupersymmetryDemystifiedMcGraw-Hillp14ISBN 978-0-07-163641-4.
36. SWeinberg (1995)The Quantum Theory of Fields1Cambridge University PressISBN 0-521-55001-7.
37. RPenrose (2005)The Road to RealityVintage Bookspp437, 566–569ISBN 978-0-09-944068-0Note: Some authors, including Penrose, use Latin letters in this definition, even though it is conventional to use Greek indices for vectors and tensors in spacetime.
38. L.HRyder (1996)Quantum Field Theory (2nd ed.)Cambridge University Pressp62ISBN 0-521-47814-6.

39. S.MTroshin; N.ETyurin (1994)Spin phenomena in particle interactionsWorld ScientificISBN 981-02-1692-0.
40. C.WMisner, K.SThorne, J.AWheelerGravitationp1146ISBN 0-7167-0344-0.
41. ICiufolini; R.R.AMatzner (2010)General relativity and John Archibald WheelerSpringerp329ISBN 90-481-3735-7.
42. HKroemer (2003)"The Thomas precession factor in spin–orbit interaction" (PDF) arXiv:physics/0310016 Bibcode:2004AmJPh..72...51Kdoi:10.1119/1.1615526.
43. Jackson, JD(1999)Classical Electrodynamics (3rd ed.)Wileyp548ISBN 0-471-30932-X.
44. RResnick; REisberg (1985)Quantum Physics of Atoms, Molecules, Solids, Nuclei and Particles (2nd ed.)John Wiley & Sonspp57, 114–116, 125–126, 272ISBN 978-0-471-87373-0.
45. P.WAtkins (1974)Quanta: A handbook of conceptsOxford University Presspp168–169, 176, 263, 228ISBN 0-19-855493-1.
46. K.SKrane (1988)Introductory Nuclear PhysicsJohn Wiley & Sonspp396–405ISBN 978-0-471-80553-3.
47. K.SKrane (1988)Introductory Nuclear PhysicsJohn Wiley & Sonspp361–370ISBN 978-0-471-80553-3.
48. AEinstein; BPodolsky; NRosen (1935)"Can Quantum-Mechanical Description of Physical Reality Be Considered Complete?"PhysRev47Bibcode:1935PhRv... 47..777Edoi:10.1103/PhysRev.47.777.
49. EAbers (2004)Quantum MechanicsAddison Wesleyp192ISBN 978-0-13-146100-0.
50. RPenrose (2005)The Road to RealityVintage BooksISBN 978-0-09-944068-0Chapter 23: The entangled quantum world
51. YAharonov; DBohm (1959)"Significance of electromagnetic potentials in quantum theory"Physical Review115: 485–491Bibcode:1959PhRv..115..485Adoi:10.1103/PhysRev.115.485.
52. Bell, John (1964)"On the Einstein Podolsky Rosen Paradox" (PDF)Physics1 (3): 195–200.
53. Lamb, Willis E.; Retherford, Robert C(1947)"Fine Structure of the Hydrogen Atom by a Microwave Method"Physical Review72 (3): 241–243Bibcode:1947 PhRv...72..241Ldoi:10.1103/PhysRev.72.241.
54. W.ELamb, Jr& R.CRetherford (1950)"Fine Structure of the Hydrogen AtomPart I"PhysRev79Bibcode:1950PhRv...79..549Ldoi:10.1103/PhysRev.79.549W.ELamb, Jr& R.CRetherford (1951)"Fine Structure of the Hydrogen AtomPart II"PhysRev81Bibcode:1951PhRv...81..222Ldoi:10.1103/PhysRev.81.222.W.ELamb, Jr(1952)"Fine Structure of the Hydrogen AtomIII"PhysRev85Bibcode:1952 PhRv...85..259Ldoi:10.1103/PhysRev.85.259W.ELamb, Jr& R.CRetherford (1952)"Fine Structure of the Hydrogen AtomIV"PhysRev86Bibcode:1952PhRv...86.1014Ldoi:10.1103/PhysRev.86.1014STriebwasser; E.SDayhoff & W.ELamb, Jr(1953)"Fine Structure of the Hydrogen AtomV"PhysRev89Bibcode:1953PhRv...89...98Tdoi:10.1103/PhysRev.89.98.

Index